Measuring the Data Universe

Reinhold Stahl • Patricia Staab

Measuring the Data Universe

Data Integration Using Statistical Data
and Metadata Exchange

 Springer

Reinhold Stahl
Dornburg, Germany

Patricia Staab
Frankfurt, Germany

Translation to English by Patricia Staab

ISBN 978-3-030-08342-7   ISBN 978-3-319-76989-9   (eBook)
https://doi.org/10.1007/978-3-319-76989-9

This Springer imprint is published by the registered company Springer International Publishing AG part of Springer Nature.
The registered company address is: Gewerbestrasse 11, 6330 Cham, Switzerland

Contents

Part II   The Statistics Standard SDMX

Part I
Creating Comprehensive Data Worlds
Using Standardisation

Chapter 1
Where We Stand, Where We Want to Be, and How to Get There

Abstract The data available to us all over the world are multiplying rapidly. Our fixation on these data is increasing accordingly and drives the demand for the collection of more and more granular data.

Companies are increasingly aware that they are sitting on an underestimated treasure of data. But most of it is stored in separate *data silos*. Therefore, many organisations are making major efforts to *integrate* data, to link the treasures hidden in the silos and to create a high-quality data world.

This integration requires an order system, that is a classification standard for data, to make things fit together. The international statistics community uses the data standard SDMX (Statistical Data and Metadata Exchange) intensively to define data structures for any kind of phenomena and, based on them, to develop data exchange processes, data collections and data analysis tools. We are convinced that SDMX can form the basis of a comprehensive, orderly and standardised data world in other areas as well.

1.1 Exploding Data Worlds

The data available to us all over the world are constantly and rapidly multiplying. Because the technical possibilities have grown immensely, more and more granular information—the corresponding term would be micro data, or even nano data—is being automatically recorded (e.g. via sensors). Social networks or search engines act as prominent data collectors of such micro data. Coincidently, they also drive technological developments—for example, Big Data—to deal with the volume of data generated. At the same time, about 70% of the world's population currently own a mobile phone and contribute every day to the growing mountain of data.

As more and more data become available, our fixation on them is increasing accordingly: post-game analyses of sports events have already turned into data-driven comparisons of space gain, one-on-one duel performance and percentage of ball possession. In doing so, our need for higher *granularity* (meaning the fine-grained nature of the data material) increases as if greater detail could also give us greater certainty. For instance, in the past, regional average daily temperatures were

© Springer International Publishing AG, part of Springer Nature 2018
R. Stahl, P. Staab, *Measuring the Data Universe*,
https://doi.org/10.1007/978-3-319-76989-9_1

absolutely sufficient to monitor the weather; now, however, hourly values are being recorded for individual cities or even streets.

Numbers suggest objectivity and provide a feeling of safety, and that is good. Or would we trust a pilot who, when we ask about the speed at which the aircraft is currently flying, has no other answer than "No idea, but quite fast". We fear obscurity and seek certainty; the more of it, the better. This is why we measure everything, everywhere and at any time. This is why we force the world around us—which is fluent, continuous and nuanced by nature—more and more into grids and digits.

Even when dealing with ourselves, we do not stop our "numbermania": we measure our consumed calories, our sleep duration, our pulse rate. Although, in the end, there might be only one result we really care about: Are we healthy? Did we lose weight? Of course, the business world is not spared by this trend: a growing number of large companies refer to themselves as data-driven companies—there is an increasing perception that they are sitting on a data treasure which, until now, has largely been left unused.

1.2   Gated Communities: The Data Silos

The tremendous data treasures of enterprises and institutions are mostly stored in so-called *data silos*. A data silo encapsulates the data, programs and processes as well as the information technology (IT) and professional expertise belonging to a specific field (see Fig. 1.1).

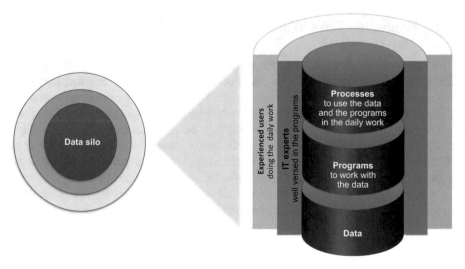

Fig. 1.1 Data silo seen from different perspectives

Data silos may be veritable treasure chests. But, just like grain silos, they seem impenetrable to the outside viewer. Grain silos can also often be underestimated, especially when looked at from a bird's-eye perspective. This is no surprise, considering that from above you see only the area covered by the base. However, once the viewer is standing on the ground in front of the silo, the considerable height and volume of the silo can be appreciated.

Data silos are mostly structures that have been developed in accordance with the actual needs of a specific department and have, over many years, been tested again and again, and ultimately optimised for regular use. Being well maintained by trained experts and developers, they offer a very high level of practicality. In addition, they are functional, robust and self-reliant; they can, for example, be set up to default to a consistent state after a system or power failure on the basis of their own data backups. Given that data silos provide such enormous value, larger companies can be expected to own a considerable, and in some cases even increasing, number of silos.

However, silos only work perfectly in isolation; the information contained within them is hardly usable outside of the silo. They use internal identifiers (IDs) or codes for products, articles, accounts, customers, suppliers and process steps. They choose their own formats for time, date, location, textual and quantitative information. Proprietary categories are created for goods, customers and territories, which in turn do not match those of other silos. All in all, if the goal was to shield the information as strictly as possible, silos are doing a fantastic job. However, this is why many companies and organisations are now making great efforts to "integrate their data": to bring the data treasures from silos into a uniform, interconnected high-quality data world. In general, the attempt is worthwhile: data integration promises high added value.

1.3   Data Linkage Is the Key

The eagerness to collect more and more granular data from more and more data silos leads to some challenges: the more fine-grained the collected material, the less valuable is the single sand grain—the piece of micro data per se. The micro piece of information is an integral part of the overall analysis and therefore needed at short notice, but ultimately it will remain only one value among many. The "useful amount of information" has therefore not grown nearly as fast as the "usable data volume". After all, hidden in these data collections lies a mountain of data points that has to be searched through.

The evaluation of a micro data set consists of suitable aggregation, outlier detection, calculation of average, minimum or maximum values, following observations over time, and so on. However, the quantum leap in the creation of knowledge occurs when micro data sets of various data silos are brought together: by linking data from different data sources, one can transform the single players into a much more powerful ensemble, as given in the examples following.

The scanners used at supermarket checkouts collect a tremendous amount of information: products and their quantities, the times and places of sales, prices, reductions and much more. A lot of conclusions can be drawn from these figures. But, of course, the information value would be even higher if other data relating to the buyer could be linked to the scanner data: name, address, age, sex, occupation, income and so on.

Imagine how big, indeed gigantic, the information value would be if one could combine the customer's supermarket data with their data from different sales points, such as pharmacies, furniture markets, petrol stations and car workshops. This is why large business corporations offer lucrative membership programmes where you collect points with each purchase and convert them into attractive reductions. In return, they collect your purchase data to create an incredibly fascinating data pool of our preferences for food, drugstore articles, prescription-free medicines, gasoline and auto-repairs. All of this, of course, with the aim of optimally tailoring their offers to our pre-calculated needs, displaying them on request and giving us personal advertising recommendations.

However, it is not only in the area of consumption that data integration represents a breakthrough in the generation of information and the development of knowledge. In the field of sciences, the linking of data from different disciplines also offers huge potential for intelligence gathering and problem solving.

Take, for example, the increasing incidence of resistant germs, which no longer react to antibiotics and have therefore become extremely dangerous. What causes the phenomenon and, more importantly, who is able to contain the threat?

- Lack of hygiene in medical facilities or places hosting massive crowds of people, such as sports stadiums? This would concern these facilities.
- Excessive or carefree administration of antibiotics for harmless diseases? This would relate to human medicine.
- Excessive or carefree administration of antibiotics in livestock farming, even as feed supplements? Then veterinary medicine and agriculture would be responsible.
- Use of expired products, potentially coming from illegal international trade? This might relate to a possible lack of working control mechanisms in this field.
- Other reasons for the phenomenon?

Examples like this clearly demonstrate that the combination of data on different phenomena can be extremely helpful for the discovery and possible solution of problems. But the same examples also illustrate the shady aspects of data integration—because in a world in which such collections of data can be created for each and every one of us, maybe even without our consent, the individual is helplessly exposed to the evaluations performed on their data, the conclusions drawn from them and, most importantly, the actions derived from them. In general, history shows us that when dealing with potentially dangerous technical advancements, ignoring their possibilities or simply prohibiting their use is not an effective response. However, the development of legal and social protection mechanisms has to keep up to speed with technical progress in order to avoid the "big brother" scenarios we fear the new possibilities of data linkage could lead to.

1.4 Data Linkage Succeeds with an Order System

To enable this vision of knowledge gain and problem solving by means of data integration to become a reality, there is a universal requirement for any "raw" data material: a good description of the data, unique identifiers for key objects (e.g. locations, products, companies) and the consistent use of uniform concepts for classification criteria or attributes (see Fig. 1.2).

To assemble the various data collections, some kind of compass or map, an operating system or classification standard is needed to make things fit together. In short, we need an order system for data. And here it shows that the information industry is, in spite of its overall high pace of innovation, in this one aspect lagging behind other branches of industry: it lacks a data standard, a *barcode of information* for the identification and organisation of intelligence. Therefore, the linkage of data from different knowledge disciplines and data sources takes very high effort to succeed.

Data must, therefore, be "put in order" to be linked: for a specific purpose, this is usually done on request by the data providers, or by the data gatherers, who must then familiarise themselves with each individual data source. The resulting standardised scheme requires a common willingness of all participants and a great deal of work, but leads to a worthwhile result: once the data are presented in such a scheme, the connection to user-friendly and powerful front-end tools that deliver "data analysis on demand" is relatively easy. They not only offer simple filtering and sorting functions for quantitative data, but with the data now linkable to other standardised data sets, it is possible to build a 360-degree view for a specific phenomenon by using various criteria (geographic aspects, age, gender, occupation, industry branch, financial indicators such as income or assets, etc.).

Fig. 1.2 Requirements for data to be evaluable

1.5   The Order System Named SDMX (Statistical Data and Metadata Exchange)

As part of our professional practice, we have worked intensively with the statistical data standard SDMX (Statistical Data and Metadata Exchange). This worldwide ISO (International Organization for Standardization) standard is commonly used in the statistical community to define data structures for any phenomena of the financial and real economy, and to develop data exchange processes and data collections as well as data analysis products based on them. Our experience in using the SMDX standard has led us to the conviction that this standard possesses the potential to form the basis for our vision of a comprehensive, organised and standardised world of data.

In a nutshell, SDMX understands every single data point, for example the "average depth of snow in alpine ski resorts", as an observation determined by several identifiers. Such identifiers could be the country where the resorts are located (using abbreviations such as AT, CH, FR, IT, DE), the year of the observation (2015, 2016, etc.), the resorts' altitude category (e.g. 1500–2000 m), the type of aggregation (average values, as opposed to minimum or maximum values) and of course the observation value itself ("depth of snow")[1]. Also, the correct interpretation of the measured values (e.g. "3.84") requires an attribute for the measuring unit ("m"). Thus, a simple SDMX structure for such a set of observations can be formed, which is able to record information on all locations, years and also other measured variables (e.g. "hours of sunshine").

Hence, SDMX provides a data model that corresponds to the *star scheme*, also called the *snowflake scheme*, which is very frequently used in IT. The actual information (called facts) is at the centre and is surrounded by the identifiers (called dimensions) in a star shape (see Fig. 1.3). Another illustration of this data model is the multi-dimensional *data cube*. In this case, the information is presented as data points in a multi-dimensional coordinate system, the axes of which are formed by the dimensions.

It is crucial to use standardised, ideally nationally or internationally agreed values—called codes—for the dimensions. Examples of codes are ISO country codes, sector codes, postal codes, global positioning system (GPS) coordinates, ISIN (International Security Identification Number) codes for securities and Alpha-ID[2] codes for medical diagnoses. This multi-dimensional data model based on concerted identifiers provides an order system and a standard that can be applied to almost all phenomena. The SDMX standard is presented in more detail in Part II of this book.

---

[1]Note that we did not include an identifier for the resort itself (e.g. the postal code), as our example contains aggregates, i.e. average values for groups of resorts. But a "resort-by-resort" data set is also possible.

[2]Identification number for diagnoses, published by the German Institute for Medical Documentation and Information.

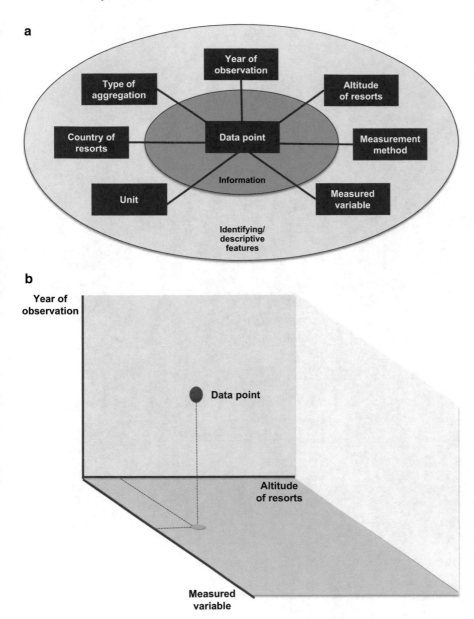

Fig. 1.3 Presentation of the information "average depth of snow in an alpine winter sports resort" as a star schema and as a data cube (for the cube, the number of dimensions was reduced to three)

So, if organisation and standardisation are so simple and promising—if the approach is so obvious—why is it so rarely applied? We look deeper into this question in the following chapters.

Chapter 2
What Does Reality Look Like?

Abstract Data are collected fervently but not wisely: often, not what is needed but what arises is gathered. Therefore, despite a tsunami of data, painful data gaps still exist.

When data sets do not fit together, the potential within them cannot be exploited. Nevertheless, the information industry has neither a system of order nor any comprehensive standard for data. This deficiency explains why firms launch countless data warehousing, business intelligence (BI) or Big Data projects, and why they appoint multiple chief data officers whose primary task is usually to bring overall order to the companies' own data.

At the end of the day, the gap will not be closed by the use, however massive, of technology alone. *Data analysis on demand* on data sets with dozens, or hundreds, of dimensions is not possible without consciousness, intelligence and expertise.

2.1 Yawning Data Gaps Despite "Collectomania"

All over the world data are eagerly, zealously, collected at every opportunity. However, there is reason to suspect that this effort is not concentrated where information is urgently needed but rather where it can be collected easily. This is why, whenever crises erupt and analyses would be helpful, data gaps are still being identified and criticised. During the 2016 refugee crisis in Germany, for example, data gaps concerning the vacancy rates of houses and flats were lamented, despite numerous data collections already existing on real estate purchases and rental prices.

Another example is the current debate on the possibly carcinogenic effect of the broad-spectrum herbicide glyphosate used for weed control. During this discussion it emerged that knowledge about the geographical distribution of diseases is incomplete. Therefore, no investigations can be carried out on correlations of (frequent) incidents of disease with the location of potential sources of danger, such as the application areas of hazardous substances in agriculture or on railway tracks, the sites of power plants or emission-intensive factories and traffic points. A similar question is whether studies on the prevalence of skin cancers could gain insight if

© Springer International Publishing AG, part of Springer Nature 2018   11
R. Stahl, P. Staab, *Measuring the Data Universe*,
https://doi.org/10.1007/978-3-319-76989-9_2

data from medicine and meteorology (solar radiation and intensity, hours of sunshine by geographic allocation, etc.) were combined.

Finally, the new international institutions for monitoring financial stability created in the aftermath of various financial crises, such as the G20 countries' Financial Stability Board, have identified a whole series of data gaps in the data sets for the financial and real economies. Currently, efforts are being taken worldwide to fill those gaps—to be better informed is to be better prepared.

2.2 The Data Universe Lacks Order

The gospel of data analysts starts with the words "In the beginning was the data". The data are the original building blocks, the atoms in our universe and the starting point of our work. They can, at the same time, be our greatest good and our most terrible curse. Because when they do not fit together, they are worthless.

Thus, the bar is set high—and the exploding data world described in this chapter is very far from this benchmark. The comparison with reality shows that the information industry lags behind other branches of industry or scientific disciplines with regard to standardisation and organisation of their most precious asset—the data. For there is neither a system of order for data and information nor a prominent standardisation, and certainly no "periodic system of elements" as in the natural sciences. Nowhere do we find anything approaching a unique identifier, a universal barcode for information. This could give the impression that Google itself is the actual system of order. This lack of data organisation is deplored within and across very different business branches or scientific fields, for example the lack of standardisation of data in genetic plant research (see Goff et al. 2011).

One might argue that this dishevelled state is to be expected given the savage and uncivilised nature of the Internet. But it also takes place in the otherwise much better managed area of industry. The lack of order is evident in the data universes of almost all companies and justifies countless initiatives: data integration projects, business intelligence (BI) projects, data warehouse projects or, lately, Big Data projects. The enormous increase in the significance of data is also reflected in the frequent appointments of Chief Information Officers (CIOs) whose main task seems to be to bring order into the company's data world.

While institutions' repeated attempts to make their own (!) data landscape manageable are usually only met with moderate success, the phenomenon of a lack of overall data organisation is even more pronounced across an entire industry branch or country. The only exceptions to the rule seem to be certain areas with specialised commercial interests—there, well-developed data worlds are available, such as search portals for used cars, hotel rooms, flight connections and apartments. However, the well-known scout websites and price comparison portals are not driven by artificial intelligence (AI). There are no AI solutions which scan through the used car advertisements of the Internet by text mining and, using Big Data technology and their intelligent networks, are able to magically determine the necessary information. No, these data were painstakingly "put into order", and thus are identifiable via a thorough classification (e.g. brand, type, year of registration, postal code of the supplier, mileage) and a complete set of attributes (e.g. presence of air conditioning, trailer coupling).

2.3   Using Information Technology Not Possible Without Content-Related Expertise

The way the new high-quality data collections formed from micro data are processed and linked has also changed significantly: given the sheer volume, diversity and complexity of the data, it is impossible to determine a priori which questions should be answered with this data material at all. This results in a much higher volatility of evaluation requirements. Information retrieval can no longer be depicted as classical, straightforward statistical production of prescribed important indicators; instead, the implementation of *data analysis on demand* is required.

For example, granular statistics on securities investments ("security-by-security") were originally designed with the objective to allow standardised, periodically recurring evaluations. However, a growing share of analysis requirements is devoted to issues that literally might have come up overnight; for example, what does the international distribution of investors in government bonds for a certain European country look like?

The linking of several data sources allows for the creation of a new style of data collections with a lot of dimensions (meaning identification features for a data point), which offer a gigantic variety of evaluation possibilities. But imagine a data cube of more than 30 dimensions: which analyst or scientist can (and would like to) handle 30 dimensions and formulate ad hoc analyses on that? In practice they will usually manage to turn three, four or five "adjusting screws" on their report (which will feel the same as juggling the corresponding number of balls), but the limit will soon be reached.

The information systems must, therefore, provide the experienced data experts with suitable data analysis products instead of dropping the expert in a labyrinth of formally fitting data that is nevertheless no longer manageable. Otherwise, there is the danger that we operate strictly within the rules but still end up comparing apples and oranges. This applies even more to the frequently used technique of data mining. Imagine this highly complex data jungle being searched through by bots, and all possible permutations of the 30 dimensions being formulated and examined for significant values or correlations of observed variables. What to make of the outcome of this technical tour de force? In this case, as well as in the case of the Big Data technique described in Chap. 3, professional expertise is of the utmost importance for evaluating and interpreting the results.

Reference

Goff SA, Vaughn M, McKay S, Lyons E, Stapleton AE, Gessler D, Matasci N, Wang L, Hanlon M, Lenards A, Muir A (2011) The iPlant collaborative: cyberinfrastructure for plant biology. Front Plant Sci 2:34. https://doi.org/10.3389/fpls.2011.00034. http://journal.frontiersin.org/article/10.3389/fpls.2011.00034/full

Chapter 3
What Can We Expect From Big Data?

Abstract The information technology industry's answer to the need for data integration is technological innovation, a current example being Big Data. Thanks to parallelisation and networking, immensely greater computing power is possible. This drives the idea of simply throwing all the data into a *data lake* and magically recovering new insights from it.

However, this brute approach soon reaches its limits—set not only by ethics but also by feasibility. For all its power, technology alone cannot solve the content-related issues of data processing and analysis; sometimes it even leads to serious mistakes. Big Data can therefore only be complementary to a scientific, intelligent approach.

Even a Big Data system needs structure, both on the input and the results side. Modelling the data with the help of a standard such as SDMX (Statistical Data and Metadata Exchange) can be used to focus and harness the overwhelming power of Big Data technology.

State-of-the-art data analysis needs to be much more dynamic ("on demand") and much faster than it is currently, while at the same time a dramatically increased data volume has to be processed. This is an almost unsolvable challenge. However, one cannot help the impression that the more frequently a challenge is classified as "almost unsolvable", the greater the number of IT providers that offer solutions for it. Countless ads, not only in relevant IT magazines but also in popular weekly magazines, send a clear message to companies: you are sitting on a data treasure and absolutely have to retrieve it—with the massive help of our IT innovations. Accordingly, IT providers suggest that without Big Data it is probably no longer possible to deal with your data. This is why we would like to make a brief plea for a reflective approach regarding technological innovations.

© Springer International Publishing AG, part of Springer Nature 2018
R. Stahl, P. Staab, *Measuring the Data Universe*,
https://doi.org/10.1007/978-3-319-76989-9_3

3.1 The Big Data Hype

Hardly any industry is equipped with such a high force of innovation as the comparatively young IT industry. Added to this is the admirable trait of constantly being ready for a new start, and leaving behind established concepts over and over again without regret in favour of even fresher ideas. It is no wonder that we find ourselves confronted with completely new and promising approaches almost every year. Gartner consultant Jackie Fenn has aptly named the phenomenon *hype cycle*.

A typical example of this is Big Data, the buzzword of the hour, which succeeded the terms business intelligence (BI) data warehouse (DWH) and online analytical processing (OLAP), which were similarly hyped in the early 2000s. Big Data systems can process distributed data volumes and accelerate data access through parallelisation and networking, and thus realise techniques that would not be possible without these enormous computing powers. One example is searching through large volumes of "unstructured data material" (e.g. text mining) to identify specific features or patterns.

At the time of writing (April 2018), a popular search engine listed approximately 62,000,000 search hits for the term Big Data; the result itself is certainly being determined using Big Data technology (the search request taking no more than 0.61 s). There is currently no IT magazine without an article and no trade fair without specific lectures on this topic. Big Data is, according to the latest survey, already being used by the most successful companies in the area of information processing. The majority of other companies name Big Data as a strategic building block for their company, as demonstrated by survey-based rankings and distributions.

Big Data, so IT promises us, will change our fundamental paradigms in dealing with data. It will replace the previous classical scientific approach—in which we first prepared a hypothesis, then obtained suitable data and finally tested our hypothesis by means of these data—with a new and much more convenient method. We do not have to think about the hypothesis any more. The tedious structuring of the data and the filtering according to relevant and irrelevant indicators is no longer necessary. No complex, well-sorted DWH has to be maintained; it takes no more than throwing all available data into a large, deep *data lake* (or swamp or sea), and from the depth of its seething centre the data-mining algorithms will, like magic, retrieve the new insights.

3.2 A Technical Approach to Big Data

Technically, the idea behind Big Data is quite simple: each computer consists essentially of the processors, the internal memory and the hard disk storage. Big Data technology creates a network of any number of computers, usually powerful server systems, to bundle their entire computing power and storage capacity and use it for a single Big Data process. This allows for almost unlimited performance and a gigantic storage volume.

For example, the entire volume of data to be processed is divided into—let's say—ten parts using ten computers, and stored on these servers, each of which then processes its dedicated data part. This may sound surprisingly simple, and thanks to the Hadoop Open Source software the basic functionality is also easily accessible; nevertheless, things very quickly become quite complex. In order for the overall process to run in a controlled manner on several computers, a type of operating system is required which controls the distribution of the data, the initiation of the subprocesses and the coordination of the results. This is essentially done by the Hadoop Distributed File System (HDFS), which controls the interaction of the individual servers (data nodes) with a higher-level management system (name node).

With this method, some types of calculation can be quite well-managed, such as, for example, counting the frequency of certain words in the parts of a divided text and merging the partial results into a total result. However, other evaluations prove to be more difficult, such as the computation of average values for the entire data set. Although the individual server does not know all the data, it can still determine the average value and a corresponding weight for its own data part. The calculation of the overall average must then take place on the higher-level name node. The more challenging the task—imagine a statistical distribution function—the more complexity has to be dealt with on the higher-level system. Already, new programming languages are being developed for this, and they have little in common with the traditional programming paradigms. They may require the use of new development techniques and lead to a new generation of IT developers.

Big Data systems require interfaces so that their results can be made available for further processing, and there are corresponding interfaces for traditional data formats or query languages, such as SQL (standard query language) or CSV (comma separated values). Also, some renowned providers of more traditional analysis applications (e.g. OLAP products) are cooperating with the Big Data framework. The technology has enormous potential for combing through very large volumes of data, analysing them for abnormalities, arranging weakly structured or even unstructured data, and preparing it for further processing in database or analysis systems. Therefore, its strengths can be demonstrated, especially at the beginning of a "data production and analysis" process. It is extremely helpful for the collection of data, and it can, if necessary, supplement or replace existing survey methods. The new technique will be very valuable, especially for the continuous brimming of data sources such as Twitter posts, search engine queries, and sensor and streaming data. However, it might fit not as well for the implementation of complex business logic.

3.3   What Big Data Is Not Able to Offer

Big Data is a miracle, a dream come true: electronic helpers, created by us, equipped with neural networks and trained by machine learning, take away the unloved tedious data work. Cinderella does not have to pick up the peas any longer; she can go straight to the ball. But the new approach has its limits. The breakthrough regarding fully automated company-wide data integration is as unlikely to be expected from Big Data as it has been realised by BI technology.

These limits are already observable in the handling of data collection. Naturally, the data are the pivotal point in any data-driven process. Ever more emphasis is being put on the technical expenses of the data collection process, and evaluations are becoming more and more attractive when the information needed for them is easily collected, i.e. as much data as possible can be obtained in the simplest possible way. The questions are thus no longer formulated along the scientific need but along the limitations of the existing data material. Supply determines demand.

Data processing also has its drawbacks: in most cases, sensible compression of the data material cannot be achieved with a simple straightforward method. For instance, outliers need to be excluded and special cases need to be treated differently, or comparisons over time only make sense when special influences have been adequately taken into account. All this requires specific skills, techniques and tools to handle mass data.

However, the real problems begin with the data evaluation. It is not for no reason that statisticians are taught very early to never establish their hypotheses nor formulate their tests ex post, that is, after viewing the data material. It is too easy for a quantitative relationship to seduce a scientist into formulating a theory that is then, sinfully, verified *by the same data*. The basic statistical principle of hypothesis before evaluation is not an end in itself—it forces the researcher to formulate individual questions consciously, objectively and honestly in advance: when is a correlation relevant? When is a level shift to be mentioned? When is the data volume sufficient to speak of the significance of a statement?

Even if the data analysis is carried out with careful consideration of all the above-mentioned effects, the interpretation of the results presents the greatest challenge. Without special professional knowledge of both content and methodology, it can lead to serious misstatements. A striking example of this is a sample survey to prove that wearing nylon stockings promotes the appearance of cellulitis. Of course, using a random sample of adults of mixed age and gender, one can easily detect a statistical correlation between the two phenomena. But are the two causally connected? Not at all! But then, how can the effect be explained? Quite simply, with the "invisible third factor": the gender of the subjects in the sample. Women, on the one hand, tend to much more often have cellulitis than men and, on the other hand, they wear nylon tights much more often. So, in the case of a sample consisting of people of both genders, there are often subjects in whom both phenomena are observed (women) or neither are (men). Hence emerges the positive correlation.

The complexity of checking hypotheses against data is also shown in the afore-mentioned example of the possible carcinogenic effect of glyphosate (Chap. 2). In fact, there are great differences of opinion between renowned organisations, includ-ing the WHO (World Health Organization), IARC (International Agency for Research on Cancer), EFSA (European Food Safety Authority) and German BfR (Federal Institute for Risk Assessment). Parts of the ongoing discussions revolve around the question if the carcinogenicity assessment should be done with the glyphosate-containing herbicides itself or with the sole glyphosate, as the active substance, alone, and in which concentration? Should it be done on the level of the pure product (which relates to the question of how dangerous any direct contact with the substance is for people active in weed control) or on the level of the resulting residues in products made of crops treated prior with glyphosate (which relates to the risk for the consumers)? In addition, the measurement of the carcinogenicity in a study caused a further conflict between scientists over the validity of experimental methods, since a mice class with a greater tendency to develop tumours could have been used. And even more general, but not less important, in this context is the question whether a genetic modification in the affected plant which leads to a resistance to glyphosate is, in itself, carcinogenic. The latter would be important if genetically modified plants were to be cultivated, imported or fed to livestock. Then things would become even more complicated:

Imagine that a pig is kept in a barn in Austria and is fed soya cultivated in Brazil, corn imported from the USA and barley grown in northern Germany. Imagine that the pig's ham ends up in a butcher's shop in Vienna, the legs as "Haxen" in Bavaria and the loin in a supermarket in Berlin. Imagine that a citizen of Berlin consumes part of it and (because of long-term consequences) falls ill with cancer 5 years later. One will hardly be able to draw conclusions on the use of glyphosate in Brazil or the USA without excluding thousands of other influencing factors.

How nice it would be if you could get rid of all these methodological problems and simply use Big Data methods to pull the "truth" out of a data lake without thinking. It would be nice, yes, too nice—and there lies the even greater danger: namely in the temptation to choose only those data sources and algorithms which confirm the favoured conclusion. Then it would be nearly impossible to conduct a methodological discussion, as in the aforementioned example about a potentially wrong class of mice being used. It would be a reversal of the high principle of mathematics, which has been taught for generations, "given are the rules, the result is sought" into a principle of "given is the result, an algorithm is sought which spits out exactly this desired result".

3.4   Ethical Concerns

The automated evaluation of large masses of data is, therefore, in principle to be considered with caution. Not for nothing, the social researcher Danah Boyd (Boyd and Crawford 2012) postulated major warnings about this approach in 2012. In addition to general ethical concerns ("Just because it is accessible doesn't mean using it is ethical"), she admonished analysts to stay unimpressed by the sheer size of the data material ("Bigger data are not always better data"), and warned them against falling for the apparent incontestability associated with quantitative analyses ("Claims to objectivity and accuracy are misleading") or neglecting the task of meaningful interpretation ("Taken out of context, Big Data loses its meaning").

However, past and present successes as well as the considerable impact of IT innovation sometimes tempt IT service providers to formulate substantial statements about data content rather than staying in their field of core competence, namely the technical tasks. Yes, Big Data technology is very powerful. Yes, it can lead to astonishing additional insights. But to believe that long-established laws for responsible information evaluation could be overridden by sheer masses—following the slogan "If I only have enough data, it will work!"—would be not only naïve but even dangerous.

On the one hand, it would be naïve to take the advertising messages of the IT marketing machinery at face value. "Schema on write", they are saying, will soon be replaced by "schema on read", which means that the data does not need to be structured when loading it into the system, but the system itself will structure it during the evaluation. In reality, it means nothing but the fact that the data is stored in its original structure. The conversion of the data into the structure relevant for the respective evaluation has to be performed on demand, based on rules given and configured by the data analyst. If the input structure of the data should change, this would not require a cost-intensive program adaptation but merely a conversion of the rule set for the evaluation. Thus, the software becomes more flexible as it can be parameterised. However, these parameters must still be deliberately written and fed in, and these rules must be consciously maintained by the data analyst.

On the other hand, this belief in the sheer mass of data would be dangerous because, for multiple reasons already mentioned, no binding statements should ever be made from purely quantitative effects. Interpreting data correlations requires intelligence, and the assumption that the machines already possess this kind of intelligence is—to date—still wrong. Consciousness, the gift of thinking beyond a single, however complex, algorithm, is for the moment still reserved to "bio-brains"—human beings. In this regard, Big Data can and should only complement a holistic scientific approach. It can serve as inspiration and impetus for investigating new contexts. It can enable us to test out new ideas, approaches and methodologies on even larger, more complex data structures. But it should not make us close our eyes and abandon our power of logical conclusion in favour of number-driven algorithms.

3.5   SDMX and Big Data: Complementary, Not Contradictory

The incomparably stronger and faster Big Data systems will be indispensable for dealing with the big bang-like expanding data world. After all, it is highly unrealistic that all new emerging data sources (sensors, social media accounts) will spit out data that are well-ordered and uniformly classified from the beginning. Therefore, techniques are needed to deal with this wild rate of growth. But even these systems cannot work miracles. Still, everything an IT system—in this case an analytical service—is to do for us must be explained and taught to it beforehand. In order to recognise patterns in material, the system has to learn the material's components, i.e. the raw matter with which it is to work—the data. Ordering the data by modelling it with a standard such as SDMX can provide the missing piece to focus the unbridled force of a Big Data technology—to bring its power on the road.

Imagine determining the consumer acceptance of a specific product with the help of a Big Data process that combs through very large texts in order to find the word "beautiful" in the comments on the product. Here, too, the process must be trained to recognise comments such as "beautiful", "very beautiful" or "extraordinary beauty" as positive and to sum them up correspondingly. But terms such as "not beautiful", "the opposite of beautiful" or "everything but beautiful" must be considered negative and scored accordingly. And we have not yet touched upon the challenge of interpreting linguistic variants such as irony or sarcasm ("If this is beautiful, I am Mister World"). Nuances such as this cannot be handled without human intelligence.

It remains that the beginning of every data analysis is the understanding and ordering of the data. Thus, in the example given, the structure of our language represents the order system that has to be understood to interpret the displayed speech elements. To order data means to understand data. And only those who understand data can interpret the messages contained in them. However, one should not expect rapid success: it takes time to establish an order system. Therefore, our efforts cannot tame or restrict the massive flood of data, only follow it. But, as in the event of a flood, while you have to take immediate care of the huge masses of water, you still have to stick to a long-term plan and preemptive measures to establish intelligent controls, such as polders, channels or corrections of river flows.

The valuable addition the SDMX order system provides to Big Data techniques is particularly evident when it comes to using the results. After all, to where and how is a Big Data process to give its findings and results? As described in Sect. 3.2, interfaces to connect the Big Data technology with analysis software products (e.g. OLAP systems of well-known IT companies) have to be provided. These, however, require the Big Data process to deliver structured data. So, the use of state-of-the-art technology (Big Data) and the use of an order system (SDMX) are perfectly complementary in their contributions to the added value of information mining, processing and analysis (see Fig. 3.1). For example, Big Data techniques are used to scrape web data sources for real estate and rental price information. However, the results are only usable in an ordered data world, classified by cities, city

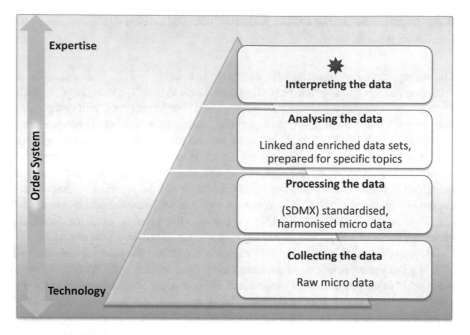

Fig. 3.1 Added-value pyramid of data analysis

parts, streets, apartment types and sizes, age, interior and equipment and so on. SDMX provides the ideal basis for this result output.

Big Data and SDMX can also ideally complement each other on a technical level. After all, the combination of many SDMX data sets is nothing more than a giant data lake, and the use of Big Data techniques for the search within this lake would make perfect sense, for example via a "designed-for-SDMX" add-on to Hadoop systems.

Reference

Boyd D, Crawford K (2012) Critical questions for big data: provocations for a cultural, technological, and scholarly phenomenon. Inform Commun Soc 15:662–679. http://www.tandfonline.com/doi/abs/10.1080/1369118X.2012.678878

Chapter 4
Why Is Data Integration So Hard?

Abstract Data integration is a multi-step process, starting with the *logical centralisation* of data in a common system. Then, a common order system, a unified data modelling method, is established to enable automated handling of the data. A common understanding will then be achieved through semantic harmonisation.

Data integration allows linking and subsequent processing of data from different sources. Essential for this is this three-step standardisation, which unfortunately has to overcome many obstacles. Some lie in the technology, some in the silo thinking of different parties involved. It may be concerns regarding privacy, or the lack of incentive due to seemingly low market potential. As a result, previous information technology standards for data were either industry-specific silo solutions or limited to a formal framework.

Why is there no universal order system for data? Creating order from chaos is a task that most people can only grudgingly bring themselves to shoulder. Even in childhood, cleaning up one's room is one of the least-loved tasks, and most people keep this instinctive dislike until they become adults. Why this is so may forever remain a mystery. Upon a closer look, we find that "creating order" is a two-step process: Step 1, set up an appropriate ordering system; and Step 2, apply this system to the area to be ordered. This sequence is unalterable, with the two steps always to be found, whether they apply to contents of a student's pencil case or to the literature references of a doctoral thesis. Step 1 is always the hardest. Transferred to the world of data, Step 1 means the development of a standard for data classification. And that is exactly where the problems start.

But let's take a step back first—let's try to describe more carefully the frequently used, but surprisingly complex, term "data integration", and let's try to explore the essential role the order system plays in it.

4.1   What Is Data Integration?

From a user's point of view, data integration might perhaps be described as the process required to obtain actual knowledge from a wide range of information (see Fig. 4.1). But what exactly is happening during this process?

The general concept of integration means embedding something new and foreign into an existing environment. This might imply adapting it to the system's "culture", which has already been defined via rules, laws, degrees of freedom, commitments, requirements and so on. This idea can be transferred to the world of data processing.

The process of data integration can be described by means of a step ladder (see Fig. 4.2). The first step is the *logical centralisation* of the data. This is the act of storing the data in one common system. It does not matter whether the data are physically (i.e. actually) or virtually (e.g. by linking) part of this system. However, it is important that the system offers a central entry point, from which all users get uniform access to the data.

Data integration is, however, more than just the collection of data, just as in real life integration means more than a coexistence of mutually exclusive subcultures in society. At this point in the process, there is still a lack of mutual reference between the data sets. Therefore, integration, as a second step, requires a common system of order; more precisely, a *uniform data modelling method*. It is only through the use of a uniform language—general concepts and terms for the way we think and talk about data—that rule-based (and thus automatable) treatment of the data becomes possible. The SDMX approach delivers such a uniform language, in the description and classification of data by "dimensions", "attributes", "concepts", "codelists" and so on.

The actual integration in the sense described, however, requires a third and final step: *semantic harmonisation* in the understanding of the content. This is the only way to make sure that the information modelled with the help of the dimensions,

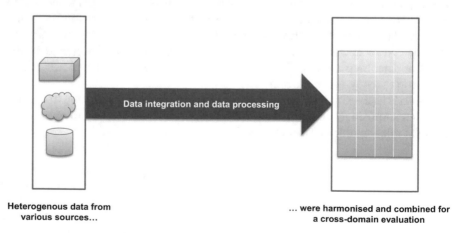

Heterogenous data from
various sources...

... were harmonised and combined for
a cross-domain evaluation

Fig. 4.1   Data integration from a user's perspective

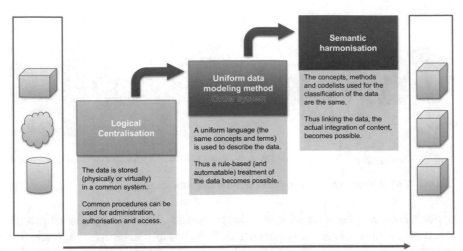

Fig. 4.2 Step ladder of data integration

attributes and codelists is essentially the same. Therefore, it is important to establish and maintain a uniform meaning of terms such as "credit", "price" or "amount". For example, it does not suffice to rule that the concept "height" is used to specify the height in metres with two decimal places. Still, it is necessary to know whether it is an altitude (height of a place above sea level), the height of a building (above the ground level), the height of a floor in a tenement house or the height of the jumping board above the water level of a swimming pool. For this purpose the concept "height" has to be completed by a textual explanation of its understanding. Modern data communities try to reach semantic harmonisation via the introduction of dictionaries, ontologies, methodologies, repositories and so on.

Why all this effort? These steps of data integration are necessary to enable a rule-based (and thus automated) combination of different data from multiple data sources into a high-quality, interdisciplinary collection of data. In other words, returning to our original example in Sect. 3.1 on the possible carcinogenic effect of substances, an integrated database of medical, geographic, meteorological and administrative data would make it possible to track down the unknown correlations of spatial disease clusters with the locations of potential hazards.

The actual handling of such a huge and multi-dimensional data collection is, in itself, very complex, and thus the trick is to "link and then simplify" the data. Because when several smaller data cubes are compiled, semantically matched and then linked, what emerges is a "super cube", i.e. a gigantic cube with many dimensions and an enormous variety of content, which actually cannot be operated by a human being. Imagine the many dimensions of the aforementioned collection of medical, geographic, meteorological and administrative data. In order to tackle the question, for example, of whether there are disease clusters in the vicinity of the

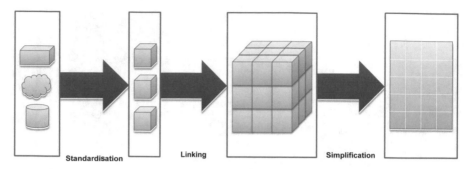

Fig. 4.3 Complete process chain of data integration and data processing

application areas of a potentially hazardous substance, usually only a small part of
the data has to be taken into consideration. The super cube must therefore be
simplified in the interest of operability and, so to speak, "flattened",
i.e. transformed into a tabular, understandable structure. Once this structure, which
is stripped of unnecessary information, has been established, easily understandable
and user-friendly analysis products, for example so-called *dashboards*, can then be
created for the user. Figure 4.3 describes the entire process of standardisation,
linking and subsequent simplification, which ultimately leads to the result that the
data users expect from data integration.

Every step in this process chain can be technically supported, and possibly
automated. One could therefore create a "production line of data integration and
processing". However, for a complete data integration process the above-mentioned
standardisation steps are indispensable: first the logical centralisation, second the
uniform data modelling method, and third the semantic harmonisation. The first step
has become a solvable task thanks to modern IT innovation. The third step involves
an intensive professional debate which cannot be undertaken by anyone but the data
experts. The second step, the consistent use of a uniform data modelling method—an
order system—forms the link between technical and specialist standardisation. In
our subsequent considerations on the challenges for the implementation of data
integration, we focus on this second step.

4.2 Innovation Speed of IT

The official establishment of a new standard usually requires a considerable amount
of time, because a large number of national and international coordination processes
are to be carried out in the corresponding standardisation or legal bodies (e.g. ISO),
before it is approved. This also applies to IT standards. The establishment of the
SDMX statistical standard as an ISO standard took approximately 4 years, although
no substantial changes were made to the definition itself.

The development of corresponding IT products is, of course, carried out at a much higher speed, so that the official publication of a standard often has to function as a sort of ex post standardisation. This is not unusual. For example, at the time of writing the European Union is still trying to introduce a standard charging device for all cell phones in 2017, but this is unfortunately years after each manufacturer created their own interface (and several years after their first attempt in 2009, when a major proportion of smartphone manufacturers "voluntarily committed themselves" to offering unified charging cables). The world of IT, in contrast, is used to the meteoric development of new platforms and methods. This does not come as a surprise: the business of software engineering allows for and requires high development rates and very fast time-to-market product releases.

This also explains the rapid rise of the World Wide Web. The big bang-like emergence of the Internet phenomenon clearly even took politicians by surprise, because they "forgot" to introduce an "Internet tax" or an "e-mail fee", whereas almost every other good is taxed. (We do not, of course, want to exclude the possibility of deliberately waiving taxation in order not to hinder rapid growth; unfathomable are the ways of politicians.) In Germany, the Federal Network Agency publicly auctioned licenses of UMTS (Universal Mobile Telecommunications System) frequency blocks to licensed mobile telephony providers. Gigantic sums of money were paid. However, the question remains open whether these auctions accelerated or hindered development.

On the other hand, the rapid technical development of the World Wide Web was only possible because the industry agreed at a very early stage on the use of the http(s) protocol and HTML (hypertext markup language) as de facto standards. Once these two definitions were imposed, implementing a web browser became a relatively simple exercise. This, in turn, explains the rapid emergence of a huge variety of web browsers.

To conclude, the world of IT and Internet, not unjustly called "Neuland" (uncharted territory) by the German Chancellor Angela Merkel, is being conquered with pioneering speed—too fast to take care of more than the basic necessities, and certainly too fast to sort and standardise it thoroughly. So, until now, there simply was not the time for a universal data classification standard.

4.3   Competition of IT Manufacturers, Products and Ideas

As already mentioned, IT product development is done very fast; the last two decades served as a downright start-up period for software companies. As a result, ingenious ideas for new products emerged on a daily basis and frequent IT system changes were made, meaning less focus could be placed on the establishment of a consistent data foundation. Still, there were certain standardisation efforts. For example, there is the query language SQL for relational databases; however, most manufacturers had their reasons for developing their own SQL derivative containing proprietary language elements (e.g. stored procedures). And, even though relational databases are well-established and provide a reliable common ground of function-alities, there are always innovations, for example NoSQL systems.

A similar phenomenon can be observed regarding programming languages. Compiler building is obviously a discipline favoured by many computer science students; correspondingly, there are always new programming languages emerging, and we are still far from a universal standard language. The combination of enor-mous creativity and competitive pressure is what drives the rapid development of the IT domain and is therefore assessed positively by us. But it also means that, as we concluded in Sect. 4.2, there simply were neither the people nor the time to define a universal data classification standard.

4.4   IT Projects Instead of Business Projects

One side effect of the rapid development of IT during the last two decades was that IT customers changed their behaviour—all of a sudden the innovation initiatives of companies were undertaken as IT projects rather than business projects. The slogan fitting this IT-dominated approach best is probably "IT as enabler"—enabling meaning putting the business areas in a new, more powerful position. Underlying this was the conviction of many companies that stronger, more expandable and ultimately more competitive work processes could only be established when the most modern IT equipment was used.

We are not calling into question the opinion expressed in this statement. How-ever, we observe that a more or less subconscious race for the use of the latest IT technology has been established. Nowadays enterprise IT architecture decisions seem to be driven by the pressure to introduce a certain technical innovation simply because competitors have already done so or because this technology promises certain general performance leaps. Therefore, the focus moved away from a solid actual-target analysis of the company's current work processes and the question of which exact improvement the affected technology could offer there. Encapsulating this phenomenon is the quote from the American behavioural psychologist Dan Ariely: "Big Data is like teenage sex: everyone talks about it, nobody really knows how to do it, everyone thinks everyone else is doing it, so everyone claims they are doing it" (Ariely 2013).

But do these activities result in the desired objectives? For several years, a phenomenon called the "productivity paradox of information technology" has been a topic of discussion. It is also called the Solow paradox, as it was introduced by the American economist and Nobel laureate Robert Solow. He described this paradox in a 1987 review of the book *The Myth of the Post-Industrial Economy* by Stephen Cohen and John Zysman (Solow 1987). The basic message of the paradox, though not sufficiently proved, is that in several business domains a high investment in IT does not lead to increases in productivity and profitability.

This statement confirms our suspicion that IT investments are often made pri-marily in order to use the latest technology, whereas less attention is paid to optimally combining the IT and business processes. The ideal balance to be sought and pursued lies in utilising technical progress while focusing on concrete and achievable benefits.

This is precisely where a data classification project comes in. In a scenario such as the above-described strive for innovation, such an endeavour does not represent any visible technical progress and has therefore generally never been the focus of attention. But, as described in Sect. 3.5, it could provide the ideal complement to a technical expansion in order to exploit the possibilities of new technologies for purposeful use.

4.5   Individualistic Mentality

In a simple way one could say that "clever minds stand in the way of standardisation". In fact, highly creative and productive personalities tend to be harder to inspire for an idea that leads to a (partial) reduction of individual freedom and at the same time to an increased burden. After all, keeping things in order always means a bit of effort. For example, some software developers like to set standards or write reusable code, but they themselves do not like to reuse the code of others. Also, in other areas of knowledge, as far as competitive pressure allows, individual products offer a stronger personal identification for their engineer. They lead to a higher degree of emotional involvement and foster self-perception as artists or creators.

In fact, proprietary solutions can generally be built a little faster and more directly than the, often burdensome, implementation based on a unified data classification system. On the other hand, when creating an individual solution it is often overlooked that a lot of effort must be put into building functionality that the standard already offers. It has to be noted that, similar to the above-mentioned positive aspects of competitive pressure, the individualistic mentality of software developers is of great value and has made many innovations possible. Nevertheless, it can be said that there is a tendency not to use potential standards for as long as possible. This reluctance usually only vanishes after a de facto standard has already been clearly established.

Once the initial resistance has been overcome, the possibilities offered by the new standard are usually very quickly appreciated. Ultimately, the standard ends up being recognised as a useful extension of one's own toolbox. Thus, the potential of the SDMX standard propagated in this book, which we briefly explained in Sect. 1.5, and which we present in detail in Part II of this book. Regarding room for creativity, SDMX only provides a framework for data management, leaving infinite freedom for the design of end-user products that clearly show the handwriting of the developers.

4.6 Silo Thinking Rather Than Interdisciplinary Thinking

Further obstacles to the establishment of standardised data collections are the defence mechanisms of the established predecessor systems—the aforementioned data silos. In nearly every company, silo thinking is denounced as standing in the way of the efforts to integrate data and as impeding the process to build a cross-domain knowledge base. However, we do not see silos as such a bad thing. Data silos, as well as their associated processes and products, have one argument to speak for them: they function properly.

In addition, the creators and operators of data silos are highly experienced and competent regarding this specific combination of technology and business cases, which ensures that the silo will continue to function in the future. The unique position of these specialists naturally entails an expert status—a certain degree of seniority. So, it is only natural that the demand for overcoming silo structures leads to concerns about the loss of a unique selling point. The same applies for transparency as well as change in data ownership (i.e. sovereignty over the interpretation of one's data), which is often feared for the same reasons. In this respect, one can observe the psychology of power in all its facets and effects because data is power, and the one who owns the data has the power.

Besides, the supporters of silos are not without argument: the implementation of a large, powerful and stable product is an enormous achievement, which often takes a long time—as opposed to the very short time it takes to tear it down. This is why in many cases there is also a justified concern regarding whether a more generic product will be able to match the functional capability of the established silo solution. Eventually, there is always the global consensus: "We have to overcome the constraints of silos and arrive at cross-domain solutions. For my silo, however, this will not be possible because … " Please feel free to fill in any one of an enormous number of reasons.

We believe the solution to this conflict should not be an either–or outcome, but rather a coexistence. The functions and processes of the silos are required as operational procedures. Most of the time they do not have to be replaced in order to build a common knowledge base for analysis—a transfer of the cross-domain-relevant data and results in the form of quality-proven *clean copies* to a universally standardised data hub is usually sufficient.

4.7 Privacy and Data Protection

One of the central purposes for ordering and standardising data is, of course, to be able to link data from different sources and to integrate different data sets into a central model. And here lies another challenge: data protection requirements.

Of course, there is always the basic question of which data may be placed into a uniform data collection, and who will be allowed, as well as who is professionally competent enough, to read the data, interpret it, link it with other data and make meaningful interpretations on the basis of the data. Data can be confidential for very different reasons: the data might contain personal information, market information covered by compliance policies, business secrets, unsecured findings, premature and sometimes unpleasant things. There are, therefore, many reasons to oppose giving others access to it.

For this reason, *confidentiality* is an extremely important aspect of statistics, considering statistics is the scientific discipline of gaining information through data. It is a *conditio sine qua non*, an essential principle. The collection of high-quality data is not possible without the trust of those who provide the data. Anyone who contributes to a statistical survey must be absolutely sure that the information given by them is not used against them. This is why there are numerous national and international statutory regulations which are, of course, taken into account in all official statistics and which must also be followed when establishing integrated data worlds. In particular, this applies to the protection of personal data. Here, further processing is ultimately only possible when the data are effectively anonymised.

The sensitivity to data protection is, partly for historical reasons, particularly high in Germany. There, recent surveys have shown that people value data protection most highly when it comes to their personal financial and health data, as well as their personal identifiers (e.g. tax identifiers, social security numbers, fingerprints and places of residence). And here lies the actual dilemma. We all hope for profound progress in in the areas of security, health and finance. This progress requires knowledge building and is not possible without data. But in this exact domain, we are also most sensitive to the transfer of personal information to public or private institutions.

On the other hand, we are more than willing to hand our personal information over to social media when it comes to curriculum vitaes, personal interests and even photos. Doesn't it come in handy when our search engine gives us useful tips (e.g. filling stations, shops) relating to the place we're at? All we have to do is give them free access to our current location...

Statutory regulations represent our cultural consensus on a specific question; this also applies for data protection laws. It is absolutely vital that the rules are observed when data are integrated; confidentiality, for example, must be ensured through effective *anonymisation* techniques and access rules. But the law does not prohibit the realisation of an ordered collection of data, per se, as the basis for data linkage and knowledge building.

4.8 Lack of Direct Incentives for Data Providers

A huge part of the general data needed for building a high-quality data world falls in the responsibility of (semi-) public institutions. Examples are weather services, health authorities, tax authorities, employment agencies, land registry offices or statistical entities. These institutions tend to publish their data on their websites, often in the form of tables or charts, and rarely as a service for direct use in data-processing software. They themselves profit little from a reciprocal coordination of their data and metadata, apart from the singular scenarios when others' data can be used for quality management of one's own data. There is neither a mandate nor funding for cross-institutional investigations. An exception to this rule is the provision of data for research purposes, which is often through research data centres funded by public money.

Apart from these special promotions, the linkage and integration of semi-public data currently does not seem to offer any kind of instant gratification. The real problem seems to be that although data are the oil of the twenty-first century (Sondergaard 2011), no one wants to pay for its production and refining. But could it not be possible that a highly profitable investment opportunity is hidden here?

Counter-models to the public world are those scenarios where users voluntarily provide their data, either as part of business terms or because they are promised a lucrative consideration. Thus, the data are produced by the clients themselves. Common examples are social media platforms such as Facebook or Twitter, online market places such as Amazon, as well as frequent-flyer miles and discount cards. The area of data to which these models refer is, of course, domain specific, so once again we are talking about silos.

4.9   Insufficient IT Standards for Data

Data in itself, in its abstract form, is not a concrete product for an end user, in contrast to music files, videos or apps. Data offer only the bare information; they are but a building block for a high-quality product, like cartographic data is but an ingredient for today's indispensable navigation systems. Since there is no immediate universal use for them, the idea of data standardisation does not prevail so easily. It is commonly accepted, in contrast, that in order to make the data visible or usable there have to be domain-proprietary expert systems. This is the only explanation for the fact that previous data-related standards were, unfortunately, either branch-specific silo solutions or formal frameworks, e.g. relating to file formats.

Of course, several industry branches defined proprietary standards for their data at an early stage. For example, EDIFACT (Electronic Data Interchange for Administration, Commerce and Transport) data formats were introduced a long time ago in the automotive industry. EDIFACT implemented the ideas of markup languages for data and of self-explanatory data streams; thus, it was a predecessor of XML (eXtended Markup Language), but one which lost importance when XML began its triumphant sweep through the Internet. Such industry-specific standards represent a wonderful example of non-matching silos.

With the cross-domain establishment of XML, industry took another major step in the right direction. Through an XML schema[1], data are given a structure; data becomes understandable, almost self-explanatory, when marked by XML tags. The appropriate term for this is *machine-readable*. Some would even say *human-readable*, although this would be a long stretch. A lot of common software products such as the popular browsers know the language of XML files and can visualise the data they contain.

XML was an important step towards an order system for the data universe: for the first time, it made it possible to send a self-explanatory data stream, the structure of which did not have to be known beforehand by the recipient in order to be able to receive and visualise the data. Some XML enthusiasts therefore assumed that XML itself would be the order system the data world needed. Sadly, this is not the case.

The demands of data processing go much further than just visualisation. Data are to be imported and handled further by automated processes. They are to be evaluated, aggregated, calculated and linked to each other according to different criteria. These requirements and the data growth described earlier in this book result in an urgent need for a standard dedicated specifically to data, a standard which enables the industrialisation of information processing, where the data fit in like building blocks.

To make this possible, this standard must do more than XML—it must also cover the semantic aspects of the data set. It must take the step from a purely formal description of the data to a classifying description of the content. This includes the separation of facts and descriptive dimensions. And this is exactly what the statistical standard SDMX does, as described by means of the example in Sect. 1.5.

XML with its formal framework thus provides an important prerequisite for the desired order system, but can only function as a carrier system for another standard to be formulated in the XML language. This is the principle of the XML-based versions of the SDMX standard, as we explain in more detail in Part II of this book.

References

Ariely D (2013) Facebook post on 1 Jan 2013. https://www.facebook.com/DanArielyOfficial/. Accessed 20 Feb 2017
Solow R (1987) We'd better watch out. New York Times Book Review, 12 Jul 1987, p 36
Sondergaard P (2011) Gartner says worldwide enterprise IT spending to reach $2.7 trillion in 2012. Analysts discuss key issues facing the IT industry during Gartner Symposium/ITxpo 2011, October 16–20, in Orlando. http://www.gartner.com/newsroom/id/1824919. Accessed 20 Feb 2017

---

[1]The XML language provides a concept for the formal specification of the elements in an XML document: XSD (XML schema definition).

Chapter 5
Basic Thoughts About Standardisation

Abstract Standards tend to be inconvenient: they replace existing proprietary solutions and cause huge migration efforts. They show their value only after these initial investments have been made. Often they are not even the "best solution" to the single individual problem. However, the strength of a standard does not come from its genius but from the fact that it is taken up by all. Once a standard has been established or even endorsed by official authorities, it is almost unstoppable. It creates reliable interfaces and promotes decentralised work. Extensive data and process standards enable the development of completely new approaches, such as blockchain technology. In our view, standardisation has the potential to trigger comprehensive and targeted data usage both within and beyond companies.

After the preceding long list of obstacles, challenges, moral dilemmas and conflicts of interest, it is finally time to make our plea *for* standardisation. In our view, standardisation can activate the true potential for cross-domain, target-oriented usage of data both within companies and beyond. But first let us share a few basic thoughts.

5.1 Standards Do Not Fall from the Sky

Two universally valid wisdoms exist regarding standardisation:

1. It is always too late for it.
2. It is never too late for it.

The first statement is correct, because a new standard almost never enters a previously empty area, but rather something always will be replaced by it. In most cases, the pre-existing situation is a more or less dense jungle of proprietary solutions. Compared to these, the standard to be introduced is just another format to be considered, and the migration to this standard means extra effort. The second statement is just as true, however, because nearly every successful standard started in this predicament as the more or less unwanted newcomer and showed its impact only after initial investments had been made.

Modern Western societies celebrate the concept of diversity; individual differences are emphasised and appreciated. Maybe this is where our visceral aversion to standardising our personal environment—which includes our way of working— stems from. None of us likes to give up our "personal touch" in favour of a one-fits-all scenario, especially when it comes to one's own creations. A standard is generally accepted only after its benefits have been clearly demonstrated. This is not always easy—but it is definitely worth it.

5.2   A Standard Is Never the Local Optimum, but Probably the Global Optimum

The term *standard solution* has some quite different interpretations. In some cases, it is associated with a plain vanilla product, the simplest version of something, without any optional extras. It may work, but it misses any comfort functions and obviously also offers no fun. In other cases, the term is expressly positive: the standard ensures that everyone can work, it fits anywhere and it does not miss any important aspects. Not surprisingly, we agree with the latter evaluation. Thus, we arrive at the next basic statement about standards:

The power of a standard does not result from its brilliance, but from the fact that it is understood and taken up by everybody.

For most individual scenarios, a standard is not the optimum solution. Almost every specific business need could be covered by a purely individual solution. From an overall perspective, however, the entirety of the individual solutions will lose against the shared use of a standard. Nobody wants to get by without commonly used office formats such as Microsoft's Word and Excel. Just imagine for a second a world without any universal standard for paper sizes and the implications for printers, copiers, envelopes, staplers, file folders, bookshelves and so on.

5.3   Standards Are Accepted When They Are Usable

Regarding a specific use case, there is almost always something better than the standard solution, but nevertheless the standard prevails. In general, any solution can be established as a standard once it manages to get a significant "market share" of users on its side, so that those who do not use the (standard) solution face disadvantages. This is best done if the following rule is met:

A standard successfully gains ground when it is sufficiently simple and thus allows for a high speed of implementation and dissemination.

A messenger program such as WhatsApp is a surprisingly simple piece of software, but it managed to become a de facto standard at a sensational speed. And almost everyone uses USB (Universal Serial Bus) sticks, plays mp3 files and looks at JPEG (Joint Photographic Experts Group) files on various devices connected via WIFI or Bluetooth.

The information technology industry provides many other examples. Take the case of the so-called markup languages, which can transport format instructions in addition to the actual data content. There, the comprehensive and powerful SGML (Standard Generalized Markup Language) could not prevail against a much simpler, however less powerful competitor: HTML. HTML became the global standard for Internet websites, despite the fact that it drove web developers and web page designers nearly crazy with its rather tricky mix of native and other-language components.

If a solution affects the end-user interface, its establishment as a de facto standard automatically results in an imperative to use it for any product in this field and a virtual ban on the use of alternative solutions. Nowadays, the use of e-mails, message services and chats is obligatory; the request to send a Word file also implicitly means not to send alternative text formats (such as LaTex). Once a standard has been established or even decided by official bodies, it can hardly be stopped. In this sense, a standard, be it official or de facto, possesses enormous power.

5.4   Standards Promote Decentralised Work

Those who try to introduce standards are often suspected as pushing for centralisation. We would like to allay these concerns and offer a different proposition: it is only through standards that decentralised collaborative work is possible.

The existence of the standard USB interface ensures that random computer manufacturers can equip their PCs (personal computers) with a USB slot for which completely unrelated random USB manufacturers produce their USB sticks. The alternatives to standards in a multi-party environment are proprietary formats and bilateral agreements, ultimately leading to an extremely inefficient scenario as a whole. Often, a single or a few dominant actors evolve from this process and enforce—via market share pressure—the introduction of a de facto standard. A typical example of this is Microsoft Office (which became a standard quickly but remained proprietary for a significant time—the corresponding file formats were opened for implementation only a few years ago).

5.5   Standards for the Realisation of New Approaches— Current Example: Blockchain

Big Data may have been a buzzword for some time; however, in the context of hypes, a new kid is in town: *blockchain technology*. Some enthusiasts already attribute it with a potential even higher than Big Data to profoundly change our economic life. It is for this reason that business journalists already call it "the blockchain revolution".

This potential, which is met with hopes by some and fears by others, is based on the blockchain idea for the digitisation and direct electronic processing of any kind of business processes. Essentially, the approach works like this: business transactions are designed as bilateral contracts ("peer-to-peer") between a vendor and a customer of a given asset without including the traditional intermediaries. This allows for fast processing, great cost advantages and a higher degree of anonymity.

The blockchain community is a young, fast-growing and occasionally anarchic community of highly diverse individuals and groups of interest (see also the Box on The Blockchain Revolution). However, this community acknowledged at an early stage that the availability of sufficient standards for data as well as processes is of crucial importance for the development of blockchain technology. Standards allow for high-productivity work across the world in different aspects of a technology, they guarantee that components of different origins and mechanisms match, and they ensure that the blockchain revolution continues at a rapid pace.

This is why the ISO Technical Committee "ISO/TC 307 Blockchain and Electronic Distributed Ledger Technologies" has been in existence since 2016. Its scope is defined by the ISO as "standardisation of blockchains and distributed ledger technologies to support interoperability and data interchange among users, applications and systems".

We view this as a strong endorsement of the standardisation ideas which we share and which are presented in this book.

The Blockchain Revolution

The current momentum of the blockchain movement can be partially explained by the fact that in addition to the actual creators and drivers of the blockchain idea, many business branches are also strongly involved. The reasons for their extensive research are different: some hope to be able to eliminate intermediaries from their transactions, e.g. banks, which are currently assisted by clearing houses in securities transactions; others fear becoming the excluded intermediary themselves, e.g. there is a worry in banks of becoming obsolete in many business transactions due to the introduction of the cryptocurrency BitCoin and the corresponding peer-to-peer digital payment system.

(continued)

Even though blockchain technology is often referred to as "simple", the underlying concept is full of massive information technology (IT) innovation: at the core, there are distributed databases on a vast network of servers around the world which contain business and transaction data and constantly synchronise it among themselves.

In the layer above, there's the Distributed Ledger Technology (DLT), a building-block system for the construction of a blockchain. DLT describes how to build a consensus algorithm that can be used to test a transaction for validity. In its simplest form this algorithm only asks the following: "Does the vendor have the goods? Does the customer have the money (or rather the BitCoins)? Are there no other third-party claims?" If the consensus algorithm reaches a positive result on a sufficient number of servers, then the transaction is valid. It is then moulded into a valid contract using cryptographic methods by a "miner" and inserted into the chain of valid transactions (blockchain) as a block which can no longer be changed. This miner, or rather BitCoin miner, was selected by random procedure and is in turn paid with BitCoins for this service.

DLT also describes how so-called smart contracts are designed. Smart contracts are needed for the contractual definition of the transaction, and they are neither smart (some depict them as rather plain, hard-coded JavaScript) nor real contracts (many doubt that they are legally incontestable). However, they contain all the information concerning the transaction, and also the details, e.g. "A partial amount of the total payment is due to be paid on the day of the arrival of the goods in the port."

As a whole, the blockchain architecture is highly complex, and the mechanisms are quite hard to follow from the outside. It might therefore still appear somewhat disreputable. But it works.

Starting from the blockchain idea one can spin endless fantasies for the nearer and farther future. Ultimately, each document can be digitised, including the ownership of automobiles or real estate, and therefore any business transaction performing the ownership change of a good can be blockchained, i.e. illustrated and carried out in a blockchain manner. This might fundamentally change business life—and this is why they call it the blockchain revolution.

This could affect and interfere with many intermediaries, hence their worries. But it may also be the cause of disquiet for some citizens that their most important business issues (such as property, money, securities, vehicles, valuables) and the related business transactions are being stored in a highly encrypted and barely accessible form in some binary data cluster called blockchain. This is still far from true, but the old rule applies: "Whatever is technically feasible will at some point in time also be done." Therefore, government agencies must take this development seriously and provide

(continued)

regulation where necessary. This is not easy, because at the core the blockchain revolution is an anarchic movement that aims to allow business settlements without control or regulation by means of an intermediary—the consequent realisation of a free market.

The development remains incredibly fascinating.

Chapter 6
Standardisation and Research

Abstract Research in various scientific disciplines shows an increasing data orientation. However, researchers are not usually interested in all-encompassing standardisation but in a narrow subject area for which they need well-prepared datasets.

Technical tools can provide a high degree of automation both for data integration and for reporting. However, there is a gap between what researchers want from a dataset and what even well-ordered data structures can offer. This gap is closed by research data centres: they offer suitable linking of data from different sources, technical preparation and good documentation of the data content. Most often, they also have to ensure confidentiality.

The existence of high-potential data sources, good preparation and accessibility of data may influence the direction of the research activity: namely, research goes where good data are available.

6.1 Limited Interest in Standardisation

Recent research in various scientific disciplines has been showing a growing orientation around data. The question arises whether this momentum attributed to researchers will also contribute to the standardisation of the data worlds. But here we should not set our expectations too high. Researchers are usually interested in narrow, dedicated pieces of information and often look for pre-selected "palatable" records with the following characteristics:

- Well documented and proven quality
- Pre-selected, pre-filtered, and, if necessary, pre-aggregated
- Pre-formatted for direct use in common analysis software (e.g. tables, CSV files)
- Tailor-made because of the high volatility of requirements.

This is why researchers are less interested in the data integration and production process itself. In most cases, the desired data record cannot be generated by an automated process from a cleanly structured data world. The final stage of the data preparation usually consists of individual, partly manual, effort, which is often performed by student assistants.

6.2   Influence of Data Availability on Research

In fact, by using descriptive statistical methods on well-prepared data sets, quite astonishing analysis results can be derived. However, the statement "On these data sets good research can be done" also nurtures some doubts: does the drilling take place where oil is expected or does it just take place where it is easy to drill? In other words, do researchers look where valuable knowledge gains are expected or do they just look where the data makes it easy to look? This might not come as a surprise when one considers the nature of the research business: in the scientific world, each researcher acts as a freelancer, a one-person company, and of course has to think about production and marketing, which in this area coincides—the next paper, the next publication. In this scenario, the reusability of their data and their results for others cannot be the researchers' focus. This is common practice and ultimately ensures the freedom and impartiality of scientific work.

However, in a time when research is becoming ever more data-driven and researchers face increasing time pressure, this practice may also be in danger of becoming an instrument of influence or even control for the research activity: by controlling the availability of high-potential data sets—i.e. data sets that promise a high level of knowledge gain—it is possible to influence the direction of research and, though in a very indirect way, the research results.

6.3   Role of Research Data Centres

Technical tools can provide a high level of automation for both data integration and data reporting. This kind of technology-driven automatisation is now used extensively in the area of company reporting and/or BI.

However, even if there were willingness to use limitless resources, a gap will remain between what analysts or researchers want from a data set regarding documentation, technical preparation and usability, and what even well-arranged data structures can offer. Why? Let us take an allegedly very simple question: "Show me the balance sheets of the 300 largest cooperative banks (measured by the balance sheet total in 2012) between 2000 and 2012" (see Figs. 6.1 and 6.2).

The researcher would need to know how to recognise whether a bank belongs to the group of cooperative banks, which balance sheet position contains the balance sheet total relevant for the major context of the question, how they implement the filters and the sorting algorithms for the balance sheet total of 2012, and how to limit the data set to the desired period. Each individual question in itself might not be difficult to answer. But even the most wonderful BI software, which allows a drill-

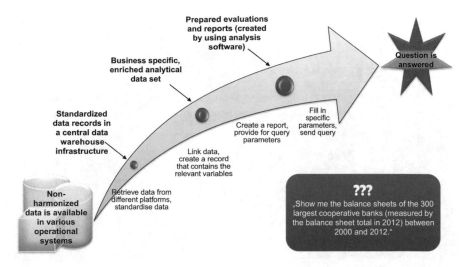

Fig. 6.1 The long way from the question to the evaluation—ideal conception

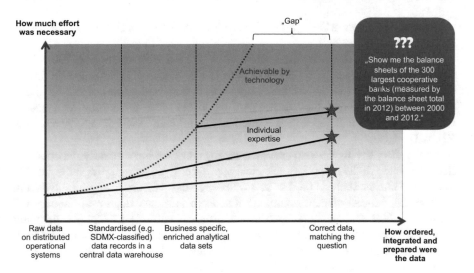

Fig. 6.2 The long way from the question to the evaluation—reality in research

down from bank groups to individual institutes, contains a powerful sorting tool, enables the confinement to specific time periods as well as a selection of individual items of a balance sheet, and can replace cryptic codes/IDs with explanatory text descriptions, suffers at least one disadvantage: very high learning costs. The researcher must therefore invest some time to learn how to implement these selections. It becomes even more difficult for him when he starts re-implementing his own algorithms for the data set in the script language associated with the BI tool. This might be the point where he gives up and demands that a data assistant provide him

Anonymisation	The data set is altered (distorted, restricted, etc.), so that the identification of data belonging to a specific individual entity would be impossible at all or only possible at a disproportionate expense.
Remote data processing	The researcher formulates the calculations without direct contact with the actual data set. Only the structure of the data set is known. The calculation is then carried out by a research data centre employee, and the results are passed on to the researcher after an output check.
Workplace arrangements (safe centres)	Researchers work on special computers that do not allow for external contacts (slots for mobile storage media, internet access, etc.). These computers are located in separate rooms, the access to which is only permitted without mobile devices or smartphones, the researchers are monitored by RDC employees.
Output checks	All results of the researcher's calculations are checked by RDC staff to ensure that they do not contain any confidential information or details.

Fig. 6.3 Examples of methods to ensure confidentiality in research data centres (Rendtel 2011)

with the selected data set as a *flat file*, i.e. a simple text file, for the analysis software he prefers and knows best.

For this reason, research data centres (RDCs) have been established by several institutions in order to close the gap in question by means of appropriate linking of data from different sources, technical preparation (e.g. in the form of flat files) and good documentation of the contents of the data. Most often they also ensure confidentiality through appropriate methods commonly used in all RDCs (see Fig. 6.3).

The staff of the RDCs also compensate for what we have already described in Sect. 2.3 as the limits of the technological approach. Even though programs are able to perform cluster analysis on data sets (cluster analysis means grouping the data in clusters, i.e. subsets of data points, so that there is higher degree of "similarity" *within* a subset as compared to *between* different subsets) to identify potential outliers and search for probabilistic correlation: all this is still purely mechanical. The ability to grasp the semantic content—the meaning—of a data set and to make the right decisions for further data processing is reserved for human beings. In other words, a machine that would be able to do just that would pass the Turing test of intelligent data processing (see Box).

The Turing Test
The Turing test, conceived by the British scientist Alan Turing, serves to determine whether a machine has human-like intelligence. Basically, it consists of a conversation between the machine and a human person (regardless if oral or via keypad and monitor). If after the conversation the person is

(continued)

convinced of having talked to another human being, then the machine has passed the test for human intelligence. So far, no computer system has fully passed the Turing test.

Reference

Rendtel U (2011) Fernrechnen, die neue Dimension des Datenzugangs? FU Berlin. 20. Wissenschaftliches Kolloquium: Micro Data Access—Internationale und nationale Perspektiven Stat. Bundesamt, 11 November 2011. https://www.destatis.de/DE/Methoden/Kolloquien/2011/Rendtel.pdf?__blob=publicationFile. Accessed 20 Feb 2017

Chapter 7
Introducing Standards Successfully

Abstract The right approach is crucial for successful data standardisation and integration. Any work with a dataset should have at its beginning the understanding of the content. Here, a *data dictionary* provides for order and serves as a base for the definition of a *data structure*.

The use of modern information technology (IT) cannot replace a sophisticated data and process model suitable for daily use. First there has to be an intelligent concept, then its realisation on an IT platform. Data integration takes time, and therefore it should be a strategic decision aimed at a long-term, evolutionary process.

Building comprehensive data worlds often means that the "data providers" have to shoulder the largest portion of the effort, whereas the direct benefits lie with other parties. This constellation calls for honesty (a clear analysis of where and how adding information to a central data collection offers added value to the company—and inclusion) a clear role concept involving all stakeholders.

Each data user community that is working on common data sets, which were produced or are still being stored in a decentralised way, has the same objective: a company-wide or even (partially) public data world that is based on a central order system. This order system is defined by classifications, data models and repositories; it is independent of products or platforms and thus represents the ideal basis for the generation of information through intelligent data linkage.

Crucial prerequisites for this are a global (in the sense of comprehensive) *data inventory*, a global *data dictionary* and—in addition to a standardised data architecture—global *coordination*. In this chapter, we give an overview of the approach with which a data community can create these prerequisites and then make the best use of their own data world.

7.1   The Correct Sequence: Start With the Content

Any work on a specific set of data should start with the understanding, the grasp, of their content. By this we mean the intellectual effort to identify the central components of the data set and to arrange it semantically. For this purpose one has to separate the actual information (facts) from the identifying or describing features (dimensions or attributes). This initial ordering step, though performed in a purely abstract way, is the right start, because it is only when a topic is really understood and intellectually incorporated into a higher-ranking system of order (data model or data dictionary) that it can be integrated into a comprehensive data world.

Only then should the technical realisation (e.g. formats, data models, databases) be addressed, without which the potential of data integration cannot unfold. Ideally, the order system does not lead to only a single solution; on the contrary, it enables internal and external developers as well as the software industry to rely on a solid basis to develop their products for the end user. Therefore, as a result of standardisation, more and more technical implementations will emerge, so that eventually the effort to integrate new data sets is reduced to a mere classification step—then standardisation achieves the desired effect.

In reality, this sequence is often not respected. All too many companies fast-forward to the technical realisation or, when proceeding with the implementation in a step-by-step approach, they plan their steps too ambitiously. We elaborate further on this in the following sections.

7.2   Creating Structure and Order

Considering the conviction that ordering the content of data is the crucial success factor, the question to ask is this: how to establish this order?

To answer this question, we take a generic view: regardless of the respective subject areas, data on all real phenomena always consist of one or more identifying features or key dimensions (e.g. licence plate number, ISIN, bank code, blood group, ISO currency code), a number of attributes (e.g. equipment of a car, methodology of a blood test, characteristics of a loan) and the actual variables measured, observed or counted (e.g. value of the balance sheet item, price of the vehicle, cholesterol value, length of an object). However, with regards to the latter, the terms "measured" and "counted" should not lead to the limiting assumption that they could only be numerical data.

For this generic approach, it is necessary to incorporate the aforementioned identifying facts, dimensions and attributes into a data dictionary, then create a data structure according to a common rule set and to document it in the data inventory. This data structure describes the arrangement of dimensions, attributes and measured variables which are used for the respective topic area; it can be described in a formal language such as an XML schema (see Fig. 7.1).

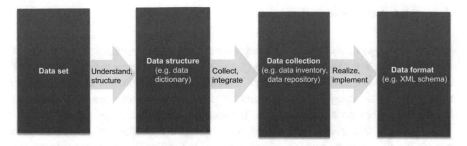

Fig. 7.1 Step-by-step classification of a data set

Based on the above-described structures, an "operating system for data processing" can be conceived at a later stage, which offers functionality for navigation, presentation, linking, transfer into analysis software products and maybe even a wizard-guided search for the data users.

7.3 Use Classification Systems and Global Identifiers

A major challenge in defining identifying features for a data structure definition lies in the non-existence of an identifier. Many areas are lacking universal identifiers. If such a central ID is missing, this can be a major showstopper for all data standardisation efforts. For example, high-quality securities statistics couldn't be developed before ISIN codes were established. And not for nothing, there are standard classification systems for the identification of diseases (ICD-10 [*International Classification of Diseases, 10th edition*] codes), company sectors (NACE [Nomenclature des Activités Économiques dans la Communauté Européenne] codes) and consumption goods (EAN [European Article Number] codes). But in many other areas such identifiers are still missing today. Urgent initiatives have been launched in these areas, such as the LEI (Legal Entity Identifier) or the UTI (Unique Transaction Identifier) or UPI (Unique Product Identifier), which are to be used for the identification of companies, their relations and business transactions.

Faced with this situation, one's own options for action are often quite restricted: one can wait for the success of an international initiative or choose the long and difficult path to introduce and maintain a proprietary internal ID. In both cases, in the short-term one will have to deal with the many shortcomings of—more or less well-maintained—"regional keys" (meaning IDs that only cover subsets of the population), which require repeated cross-matching and checking for overlaps (the sorely afflicted data specialists like to speak of "mapping"). To conclude, in most areas without a unique ID, there are not too few IDs but rather too many.

The ideal solution is, and will remain, a global identifier. So, naturally one should support its introduction and make the case for its endorsement by all parties involved.

7.4 Use Technology Wisely

In the early days of computing in companies, when the computing power of large-scale machines was measured in MIPS (million instructions per seconds), German IT engineers used to make the joke: "Grips or MIPS?" Grips being the German word for savvy, one could roughly translate this to, do we try to solve the problem in an intelligent way or do we attack it with the massive computation power of IT?

Nowadays, this question is all too often decided in favour of IT power. One reason for this is that the people responsible for the decisions are usually not IT experts themselves and tend to take the promises of IT service providers at face value. On top of that, processors and memory are comparatively cheap, which suggests that any annoying problem can be dealt with as long as the hardware and software are strong enough. Unfortunately, there are challenges that cannot be solved by massive use of time and money. Typical examples of this phenomenon are the expectations and the reactions related to the current Big Data hype described in Sect. 3.1.

The use of modern IT capabilities is absolutely essential to the success of data integration. But IT, however powerful, cannot be a substitute for a well-thought-out, suitable for daily use, model for data and processes. IT cannot replace concept; it can only follow it. The correct answer to the question "Grips or MIPS?" must therefore always be "First Grips. Then MIPS." First, there must be an intelligent design, then the implementation on one or more IT platforms will follow.

7.5 Choose Small Steps for a Step-by-Step Approach

Over the past few years, many companies have started projects to create a company-wide data warehouse. Since these initiatives have been met with all kinds of resistance—as described in detail in Chap. 4—executive power was often wielded as a last resort: "This decision comes straight from the executive board; we need a central BCC (BI competence centre) that regulates everything."

Regarding this way to proceed, according to which the highest form of motivation is force, a word of warning seems appropriate: good ideas prevail by themselves because they are inherently easy to follow. And the movement a good idea has triggered must be given a certain amount of time. When it comes to eventual success, evolution has a far better record than revolution. To put all hopes into one single big project and strive for a perfect result at the first strike might therefore be a mistake. And while we are talking about ambition, we consider it way too ambitious to demand that all (!) company data without distinction belong to the central data warehouse. This way of thinking contains the same mistake as the one we described earlier: the idea that by the massive use of IT capabilities all problems can be solved, instead of thinking about what one wants to achieve exactly.

First of all, a concept for the cross-domain data world is needed, and only then a technical solution for the implementation. Initially this would be an empty data warehouse, into which, after completion, the different topics can move in their own time and according to the actual business needs. This idea must obviously be endorsed by the highest management level of a company, since central coordination is necessary. However, this should not be in the form of a single, massively sponsored, project, which is often preferred in company practice, but rather in the form of a strategic directional decision, counting on continuous development.

But what do we actually have against the project-oriented approach? Well, projects are a fine thing, especially because they provide decision makers and auditors with verifiable objectives and a high level of controllability. But projects are always characterised by a clearly defined task in terms of time, function and budget, which is not the case when one endeavours to construct a comprehensive data world. Because even if the idea seems wonderful—create the IT infrastructure and transfer the company data to this new environment all in one great act of force—those projects nearly all fail. In the best case they are stopped at an early stage; in the worst case they are redefined and shortened only after massive expenditure, leaving scorched earth in their wake.

7.6 Treat Stakeholders Right

The power battles between the data providers and the data users, which are often observed in activities related to the construction of company-wide data warehouses, are often based on the dilemma that the data providers—the owners of the data silos—are expected to carry the burden of data integration whereas they are the ones with *no direct benefit*. By this we mean that their own work process, performed in the data silo, does not improve in any way when their information, results or data are provided, as a by-product, to others for analytical purposes. "*Cui bono?*" The people that benefit most from data integration are not the ones who have to carry out the work.

It is therefore vital to have a clear understanding of to whom the placement of information in a central data warehouse offers a real added value. The necessary work to convince people and to justify the effort can then start. The argument that simply every data set has to be integrated, because ultimately the benefit will somehow appear, does not sound very convincing. In particular, this argument suffers when the data users formulate their requirements in exactly this manner: "Give us everything, because we will soon know why we needed it in the first place."

The success factor lies in the integration of all stakeholders; this also requires a clear understanding of everybody's role. The Internet is swarming with definitions of roles and tasks which complement, overlap or even contradict each other. Typical terms are the Data Steward (who maintains the data set), Data Owner (who controls the data set), Data Expert (who knows the data set best) and Data Provider (who, well, provides the data set). And these only cover the data-giving parties, not to speak of the data-using parties.

These roles must be designed specifically for the company, all the while taking into account individual requirements for confidentiality, compliance and data protection regulations. This highly demanding process requires a sense of diplomacy and persuasiveness, but also tenacity and, last but not least, a great deal of patience.

Chapter 8
Statistics Driving Successful Data Integration

Abstract Statistics is cross-domain by nature: as a general discipline for "building knowledge through intelligent evaluation of experience" it serves as an auxiliary science in many fields. Therefore, it is statistical day-to-day business to integrate a wide range of information sources. The international statistics community can look back on a long tradition of data exchange.

Thus, statistics began to develop general terms and concepts for this purpose at an early stage. These concepts for numeric data, metadata and registry data can easily be applied to databases of various platforms. They are implemented in the statistical standard SDMX (Statistical Data and Metadata Exchange), which since 2005 has decisively contributed to the development of internationally harmonised economic statistics as well as global data exchange and data sharing.

8.1 The Cross-Domain Nature of Statistics

Why should the concepts for a comprehensive, well-organised data universe emerge from the world of statistics? Why should the necessary standards be developed here? Well, statistics in itself is a generic discipline; it is needed in almost all scientific domains. Statistics could therefore be called the general discipline for "building knowledge through intelligent evaluation of observations". It is part of the day-to-day work of statistics to reconcile the widest range of information sources—in other words, to perform data integration.

In many scientific disciplines there have been assumptions or even certainties that were known long before the relevant science could deliver an explanation for them, just because statistical correlations had been observed. Take, for example, the risk for cancer from smoking. The statistical fact of a significantly higher prevalence of lung cancer among smokers was already known in the late 1930s. It was only years later that medical evidence—that is, the biological link between the harmful substances contained in cigarettes and the development or proliferation of cancer cells—was found. The time before this was spent by smokers in a kind of residual uncertainty regarding tobacco, at least until medicine delivered definitive proof. Actually, in the 1972 (!) film "Smoking and Health: The Need to Know" (Lorillard

© Springer International Publishing AG, part of Springer Nature 2018   53
R. Stahl, P. Staab, *Measuring the Data Universe*,
https://doi.org/10.1007/978-3-319-76989-9_8

Records 1972), rather calming messages were broadcast and performed in front of hundreds of thousands of spectators. There are other examples: climate change is also statistically observable, even if in some places there are (politically or economically motivated?) doubts whether it exists at all and whether the assumptions about the causes are correct.

Particularly in areas where scientific proof is difficult, statistics are often used instead. Such is the case in psychology, for example, where statistical methods are used to prove the effectiveness of psychopharmaceuticals. Thus, statistics play a greater role in psychology than many unfortunate, mathematically challenged psychology students had imagined at the beginning of their studies.

As a whole, statistics is needed as an auxiliary science to establish or verify professional theories in various branches of science. Of course, this approach has its limits, especially when the conclusions reached do not comply with the rules of statistics. Examples of purely statistically ascertained and highly doubtful statements are abundant. For example, do vegetarians really live longer than carnivores? Or is it because, apart from renouncing meat, they also practice a more conscious and thus healthier lifestyle? Or, how significant are the ADAC (General German Automobile Club) breakdown statistics really? Once a car manufacturer succeeds in persuading their buyers to call the brand-specific hotline in the event of a breakdown, then this manufacturer will inevitably improve their results in the ADAC breakdown statistics.

In spite of the many examples of overambitious or even faulty application of statistical methods in other sciences, statistics remain indispensable in many areas; therefore, statistics is also faced with many interdisciplinary data sets.

8.2   Statistical Concepts for the Construction of a Data World

Statistical institutions all over the world have the task of collecting gigantic data sets for an equally gigantic variety of topics from all areas of life, quality-proving the data and making it available for all kinds of users, while at the same time respecting the relevant confidentiality regulations. Therefore, statistics soon started to define general terms and concepts for this task. It became clear from the beginning that two things were needed to deal with this challenge: the principle of *multi-dimensional* information and a *generic* view on data.

What does multi-dimensionality mean? As already described in Sect. 1.5, in the quick introduction to SDMX, each piece of information has a set of defining features (characteristics, dimensions, key components) to answer the most important questions: who, where, when, what, how much and so on. Every product purchase, for example, is defined by the characteristics buyer, seller, product, quantity and payment date. These characteristics—the key (components)—identify the actual information, in this case the purchase amount. This can be further described by a set of attributes (e.g. method of payment).

What does a generic view mean? It means that for the data analyst there is no such thing as micro data, macro data, financial data, supervisory data, measurement data, operational data, transactional data, accounting data, raw data or adjusted data. There are only data. And data are characterised by the previously described principle of multi-dimensionality, which can be applied to any data of a wide range of topics—generically.

Therefore, a well-constructed statistical database contains a multi-dimensional classification of the observed phenomenon with coded values for the individual dimensions, ideally taken from (internationally) coordinated codelists such as country codes or currency codes. Add to that methodological attributes, references to sources, measurement units (kg, €, m), dimensioning of the values (millions, billions, etc.), the base value for index values, type specifications (estimated, provisional, etc.), comments, and confidentiality and access rules.

At the core of such a data world lies the elementary data itself, often called variables, observations, measures or facts. These are actually most often simple *key–value pairs*, for example:

$$\text{key (bank code} + \text{date} + \text{balance sheet item)} - \text{value (amount)}$$

However, these key–value pairs are worthless without explanatory information, so-called *metadata*, i.e. data on data. The term metadata is very general and includes a variety of aspects. It might mean the structural information relevant to the automated processes for data sets, such as information about dimensions, attributes or codelists used, of a specific data collection. It might also refer to textual description and reference information on coded key components (e.g. country code DE = Germany, which is part of the European Union as well as the Eurozone, etc.). In addition, there is also an even more textual metadata type, containing extensive descriptions of the content and characteristics (e.g. methodology, quality assessment, contact details, etc.) of a data collection. These are generally not suitable for automated processing.

Micro data mostly refer to individual statistical units as feature carriers, for example the blood pressure values or cholesterol values of individual patients in a test group. When collecting micro data for individual units (persons, companies, securities, etc.), it is common to also collect additional properties (name, address, gender, etc.) for those units, which can be used for later analysis or the search for correlations. These latter data are referred to as *master data* or *reference data*. They are often kept in registers, which is why they are also called register data. These reference data are extremely important because when many different data sets refer to the same units, they may also refer to the same register, and linkage of data becomes possible. A typical example would be the use of a link between health data and financial data of individuals to examine the correlation between wealth and life expectancy.

These terms and concepts can easily be applied to data sets of various origins and characteristics. They are implemented in the statistics standard SDMX, which is described in more detail in Chap. 9. Since 2005, SDMX has played a decisive role in the creation of internationally harmonised economic statistics as well as in the establishment of international data exchange and data sharing.

8.3   Data Exchange and Data Sharing in Statistics

The international statistical community takes pride in a long tradition of intensive data exchange. Statistics is an important auxiliary discipline for many natural and social sciences, and numerous research projects could not have been carried out without sufficient empirical data. As a result, it was a common requirement for the data providers of statistics to prepare their material for a possible transfer and re-use; thus, measurement tables or data sets were exchanged already at an early stage. A wonderful historical example of a worldwide scientific cooperation on data is described by the author Andrea Wulf in her book *Chasing Venus: The Race to Measure the Heavens* (Wulf 2013): the data collection leading to the calculation of the distance between the earth and the sun (see Box).

The Transit of Venus

On 6 June 1761 and 3 June 1769, people all over the world could observe a very rare phenomenon: the transit of the planet Venus in front of the sun. Scientists had predicted that this event would take place twice in the course of 8 years and would not occur again for more than a hundred years afterwards. It offered an extremely rare opportunity, because the transits, more specifically their durations, could be used to determine a previously unknown astronomical variable for the first time: the distance from the earth to the sun. It was possible to estimate this distance by triangulation; all that was needed were several measurements of the transit durations from different points around the earth's equator. In a time that was defined by wars and conflicts for many people, hundreds of scientists set off: They embarked on daring expeditions around the globe to take these measurements at the predetermined time and locations. But just as critical for the success of the project, although less adventurous than the journeys, were the subsequent collection of these numerous data and their preparation for the calculation.

In modern times, data exchange has become much more efficient. However, it remains a burdensome task for statistics providers to repeatedly transfer and check large volumes of data. For this reason, a new concept has risen and is slowly being established: the *data hub*. In the data hub scenario, the data are provided at a central location, the hub, so that all interested parties are relieved of the effort of keeping local copies. Some call this the switch from the push-approach (where the data provider actively sends the data to all recipients) to the pull-approach (where the data users get the data from a central source when needed). One example is the Joint External Debt Statistics Hub (JEDH) of the international organisations the BIS (Bank for International Settlements), IMF (International Monetary Fund), OECD (Organisation for Economic Co-operation and Development) and World Bank.

Regardless of the form of provision (exchange or hub), the general term used when data are made available to other entities is *data sharing*; usually it is for analytical work, scientific work (research) or for a *peer review* (review of the work of another, using the same data foundation). Data sharing plays an important role in statistics, so it is no wonder that a large part of statistical standardisation effort is dedicated to this purpose.

References

Lorillard Records (1972) Smoking and health: the need to know [Parts 1–2]. https://archive.org/details/tobacco_hjy99d00.   https://www.industrydocumentslibrary.ucsf.edu/tobacco/docs/lkjy0104. Accessed 1 Dec 2016
Wulf A (2013) Chasing Venus: the race to measure the heavens. Vintage, New York. ISBN-10: 0307744604. ISBN-13: 978-0307744609

Chapter 9
Contribution of the Statistics Standard SDMX

Abstract SDMX (Statistical Data and Metadata Exchange) is a standard used successfully by the international statistics community. In data exchange, SDMX has replaced numerous individual agreements and enabled the development of so-called *data-driven systems*.

But the real potential of SDMX lies in its use as a classification system for any— and not only—financial and economic data. Many institutions use SDMX for data collection, data provision and Internet publishing. A real-life example of a mature SDMX solution is the IMF's (International Monetary Fund's) "Special Data Dissemination Standard Plus" initiative.

At its core, SDMX is amazingly simple: a data set is modelled by identifying and encoding its defining dimensions, i.e. the axes of its coordinate system. Each individual data point from this data set is then uniquely determined by its coordinates: the SDMX key. By sharing specific dimensions between different topic areas, an SDMX landscape of linked topic areas emerges.

In the previous chapters we discussed the crucial role the establishment of an order system plays in creating a comprehensive data world, the challenges of the implementation and the generic approach of statistics. Now we would like to put into concrete terms what the statistics domain has to offer in this regard.

9.1 What Is SDMX?

SDMX is an ISO standard (ISO 17369) that is successfully used by the international statistical community in many respects.

Originally, the SDMX initiative, launched by its sponsor organisations in 2001 (see Fig. 9.1), had the objective of advancing and simplifying the international exchange of statistical data. Measured against this aim, the initiative was very successful: the exchange of statistical financial and economic data between the sponsor organisations and the associated member countries, including their national statistical offices and central banks, had previously been coordinated separately by

© Springer International Publishing AG, part of Springer Nature 2018   59
R. Stahl, P. Staab, *Measuring the Data Universe*,
https://doi.org/10.1007/978-3-319-76989-9_9

| BIS (Bank for International Settlements) |
| ECB (European Central Bank) |
| Eurostat (Statistical Office of the European Union) |
| IMF (International Monetary Fund) |
| OECD (Organisation for Economic Co-operation and Development) |
| UN (United Nations) |
| World Bank (World Bank Group) |

Fig. 9.1 SDMX (Statistical Data and Metadata Exchange) sponsor organisations (SDMX 2016)

business area or statistical domain and, in the worst case, bilaterally—nowadays it is SDMX-based and standardised.

It soon became apparent, however, that SDMX was not only suitable for the mere specification of exchange formats. On the contrary, the true potential was in the underlying *information model*, which could be used to model data of any business domain. SDMX is suitable as a classification system for any financial and economic data, but also for other data, and is therefore the ideal order system for many of the institutions involved. Over time, several internal systems in these institutions have been organised as SDMX data collections. SDMX classifications and associated technical functionalities (such as SDMX web services) are also used in many places to disseminate data and publish data via the Internet.

9.2   Introduction to SDMX

The power of the SDMX approach stems, among other things, from the fact that the basic idea of SDMX is surprisingly simple: at its heart, SDMX is nothing but an implementation of the previously described principles—multi-dimensionality of information and a generic view on data.

For a subject such as the "Certainty of Snow Statistics" mentioned in Sect. 1.5 and explained in more detail in Sect. 9.3, the SDMX model is obtained by identifying the determining dimensions of the topic area—the axes of its own coordinate system. For the "Certainty of Snow Statistics" these would be the country of the ski resorts, altitude category, year of the measurement, aggregation type and type of the measured value (depth of snow) itself. In the next step, the axes are labelled: the valid values for each dimension are identified and form codelists, such as a list of ISO country keys or a list of altitude categories.

This definition is enriched by additional descriptive information—attributes such as the measuring unit "m" (metres). As a whole, the definition of the dimensions and attributes, as well as their corresponding codelists, forms the data structure definition (DSD) of the topic area. Each data set of this topic complies with the DSD. And each individual data point from these data sets is uniquely defined by its SDMX key: the coordinates on the axes defined by the DSD.

This also clearly demonstrates how important the uniform use of codes from coordinated codelists is. If in one data collection the code DE is used for the identification of the country Germany, it will not fit together with a different data collection in which the identifier DEU, GER or 0049 is used for the same country. Proprietary identifiers are probably the best solution within the specific data silo they come from—the local optimum. But if, as a means to identify a specific location, the postal code is used in one data collection, the telephone company's area code, some geo-identifiers or GPS coordinates in another, then the data just do not get along well with each other.

In the following sections, we first demonstrate a simplified example of an SDMX design (Sect. 9.3), and afterwards describe a professional SDMX solution with practical international relevance (Sect. 9.5).

9.3 How to Design an SDMX Structure: A Simple Example

In Part II of this book we explain more precisely the underlying mechanisms of SDMX. At this point, however, a simplified example is intended to illustrate the general approach. Let us take the fictitious "Certainty of Snow Statistics" (see Fig. 9.2) we have already mentioned (see Sect. 9.2).

The data table depicted in Fig. 9.2 shows several averages of the depth of snow in Alpine ski resorts, given in the unit metres. Let us now look for the classifying dimensions: in order to clearly identify a specific numerical value in the table, one should know for which *country* (AT, CH, DE, IT) and in which *year* (2015, 2016) it

Altitude		Resorts above 2,000 m		Resorts below 2,000 m	
Country	Year	2015	2016	2015	2016
AT		2.15	2.12	3.16	2.17
CH		3.16	3.16	2.17	3.19
DE		2.17	3.16	3.19	2.15
IT		3.19	2.17	2.12	3.16

"Certainty of Snow Statistics": Annual average depth of snow in different ski resorts, given in metres (arbitrary values)

Fig. 9.2 Simple statistics example: original data table

was observed, as well as the *altitude* of ski resorts (above or below 2000 m) it refers to. Our data cube would thus be three-dimensional.

In order to make the data cube expandable, i.e. able to include other measurements, we could include the additional dimension *type of aggregation*, because after all it could be possible for someone to send us not only the average but also the maximum or minimum values of snow levels next year. It is also good to introduce the *frequency* of the observation—in this case annually—as a dimension, so that the conversion to a shorter periodicity, for example monthly measurements, can be easily represented if necessary. (In this case, of course, the codelist of the *time* dimension, which currently contains only years for annual data, would have to be adapted.)

Another possible dimension would be the *unit of measurement* (metres). However, the unit is usually not defined as a dimension but as an attribute. The reason for this is rather philosophical: it is true that without knowing the unit the numerical value cannot be interpreted. However, even when switching to another unit (e.g. feet or centimetres), the true value, the essence of the observation, would not change; it would just endure a simple conversion. The SDMX DSD we designed is shown in Fig. 9.3.

Our SDMX model can easily fit in additional data points from other years or other countries; all there is to do is to extend the codelists accordingly. Also, it is possible to adapt the codelist for altitude ranges to allow a finer granularity in this dimension. The change to a higher frequency of observation has already been addressed. In the end, even the inclusion of an additional dimension that has not yet been considered—such as the distinction between alpine and cross-country skiing areas—in the model is only a small effort.

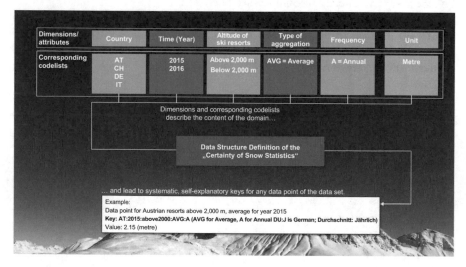

Fig. 9.3 SDMX (Statistical Data and Metadata Exchange) data structure definition for the simple data table of Fig. 9.2: composition of an SDMX key

IT:2015:*:AVG:A	Average depth of snow for all Italian ski resorts in 2015
::above2000:AVG:A	Average depth of snow for all resorts above 2,000 m
AT:*:*:AVG:A	All average depths of snow for Austrian ski resorts

(In these queries the wildcard character * is used to represent any possible code for the "wildcarded" dimension.)

Fig. 9.4 Exemplary wildcard queries on the data cube "Certainty of Snow Statistics"

One further thing: the data presented here are obviously based on (annually) recurring measurements; they are *time series* data. Since a large part of the data presented in SDMX are organised in time series, the SDMX information model contains concepts specifically designed to make working with time series easier. In most SDMX DSDs, in simplified terms, the SDMX key only describes the time series, which in turn contains the observation points (in the sample case, the observation values of the years 2015 and 2016). However, this special treatment of the time dimension is only a specialisation of the common SDMX model. In general, SDMX is also suitable for cross-sectional data or transactional data.

The benefit of the SDMX model for the corresponding topic area is obvious: SDMX ensures a uniform description of all data content, separating the concepts necessary for the unambiguous identification of a value (dimensions) from those which are only additionally describing it (attributes). The standardised structure and the use of codelists make search queries in this multi-dimensional data cube very easy (see Sect. 1.5). For our sample data cube, an example of what some search queries using the wildcard character "*" would look like is given in Fig. 9.4.

But the real benefit of the SDMX standardisation only shows itself when the SDMX model is applied to other topic areas as well. SDMX allows for the design of an individual DSD specifically developed for each topic area, so that the special characteristics of the individual domains can be included. But when the DSDs of different topic areas start using the same codelists for the dimensions that they have in common, then SDMX becomes a true instrument of standardisation. And when the different data cubes become interconnectable through these common dimensions, we are slowly advancing toward a cross-domain *SDMX landscape*. In the case of our "Certainty of Snow Statistics" example, the additional data cubes "Number of Visitors Statistics" and "Hours of Sunshine Statistics" could use the same dimensions of country, year and altitude. Linking these cubes could then help to investigate whether there is a statistical correlation between depths of snow, hours of sunshine and numbers of visitors. Of course, such an investigation requires a much higher degree of granularity and a much larger data volume.

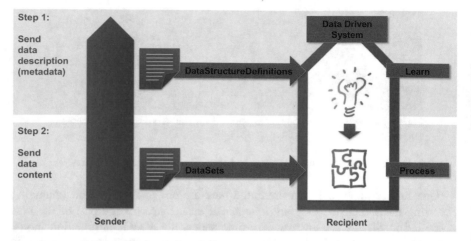

Fig. 9.5   Use of SDMX (Statistical Data and Metadata Exchange)-based systems for data exchange

9.4   Data-Driven Systems in Statistical Data Exchange Thanks to SDMX

In the world of official statistical data exchange, SDMX could successfully replace many, partly bilaterally concluded, individual agreements with a universal message format. SDMX also makes it possible for the data exchange to be based on *data-driven systems*. Data-driven systems are systems whose processes for handling data are not pre-defined and hard-coded, but are dynamically built at run-time according to the information they are receiving. In the case of SDMX, this means that the recipient's systems can process any kind of SDMX data sets they receive via the data exchange. There are only two steps necessary for the handling of a new data flow. In the first step, the SDMX DSD is communicated to the recipient, who feeds the definition to their systems. The systems "learn" the new definition and then "know" what data structure to expect. In the second step, the data contents are sent. The data-driven systems of the recipient process (e.g. validate, store and transfer) them appropriately (see Fig. 9.5).

9.5   Professional Example of Practical Relevance

An excellent example of a professional SDMX-based solution is the IMF's Special Data Dissemination Standard Plus (SDDS Plus) Initiative, which the IMF describes in its Data Collection Strategy, under the title *Leveraging SDMX Standards*, calling it "the most advanced tier of the IMF Data Standards Initiatives" (IMF 2014a).

The objectives of the SDDS standardisation, which has been promoted since the 1990s, are to increase data transparency and support the credibility of statistical systems worldwide through the use of internationally harmonised economic indicators. In response to various global financial crises, in 2012 SDDS was enhanced and the SDDS Plus added another comprehensive data collection. Basically, this supplement consisted of the inclusion of further content: economic and financial data with a view to improve information provision for financial stability analysis and crisis prevention in an environment of ongoing economic and financial globalisation.

But in addition to the enhancements regarding methodology and content, the technical implementation of SDDS Plus took a new approach, using the SDMX standard. The IMF defined the uniform global data structures to be used, which were made available in the form of a central metadata database or rather *repository* by the IMF. This central repository contains the structure definitions of the data sets to be used as well as the associated codelists for the dimensions and attributes (e.g. countries, balance of payment items), all in an SDMX syntax (SDMX DSD). The contributing countries provide their national data, which is collected by statistical offices, central banks and other institutions, on their National Summary Data Pages (NSDP). These are usually the websites of the respective national statistical offices. This is therefore a perfect example of a *central* provision of the metadata (explanatory data), combined with links to the *decentralised* provision of the data of the participating countries. Thus, one more step is taken—away from the traditional data exchange into a central data hub—towards decentralised sharing of globally distributed data. Germany's data can be found under the NSDP[1] operated by its Federal Statistical Office.

The following is an example of a time series from Germany's data on official reserve assets and other foreign currency assets. The interaction of data and metadata, which is typical for SDMX, is shown in the construction of the SDMX key (which is, so to speak, a barcode of information) and in the way the explanation of the content can be derived from the metadata (principle of self-explanatory keys).

The time series is part of the data set BBFI1 (Bundesbank International Investment Position according to BPM6[2]); its 15-dimensional key is BBFI1.M.N.DE.W1. S121.S1.LE.A.FA.R.F11._Z.XAU.M.N (the character "." is used to separate the dimension codes). The interpretation of the key is displayed on the website at the touch of a button—comparable to the handling of a barcode scanner (see Fig. 9.6).

With reference to the central metadata repository, it is easy to generate high-quality reports or graphics that show summaries or counter-representations of the participating countries in order to display the data. These interactive evaluations can be accessed under the heading of Principal Global Indicators (PGI)[3] on the website

---

[1]https://www.destatis.de/EN/FactsFigures/Indicators/ShortTermIndicators/IMF/NSDP.html

[2]The BPM (*Balance of Payments Manual*) provides international guidelines for national external statistics.

[3]"The PGI website was launched in 2009 in response to the global financial crisis, and is hosted by the IMF. It is a joint undertaking of the IAG, which was established in 2008 to coordinate statistical issues and data gaps highlighted by the global crisis and to strengthen data collection" (IMF 2014b).

No	Dimension	Code - ID	Code - Desription
1	Frequency (BBk)	M	Monthly
2	Adjustment (BBk)	N	Unadjusted figure
3	Country	DE	Germany
4	Counterpart area	W1	Rest of the World
5	Resident sector of the compiling economy	S121	Central bank
6	Counterpart sector	S1	Total economy
7	Transaction flow, position (stock), or a change in position not due to transactions	LE	Closing balance sheet/Positions/Stocks
8	Accounting entry (asset, liability, net)	A	Assets (Net Acquisition of)
9	International account item	FA	Financial account
10	Functional category	R	Reserve Assets
11	Type of financial instrument	F11	Monetary gold
12	Original maturity	_Z	Not applicable
13	Area (ISO currency codes, list of currencies)	XAU	Gold
14	Method of valuation	M	Market value
15	Compilation Methodology	N	Compilation methodology applied for national statistics

Fig. 9.6 SDMX (Statistical Data and Metadata Exchange) dimensions of a Special Data Dissemination Standard Plus (SDDS Plus) time series key (corresponding time series can be found on the German National Data Summary Page)

of the Inter-Agency Group on Financial & Economic Statistics (IAG)[4]. On this public website, a high-valuable data collection is provided in an attractive manner and free of charge; it can also be accessed via the IMF Data App (IMF 2011).

The SDDS Plus initiative demonstrates the positive effects of standardisation: the achievement of a globally harmonised classification for the most important real and financial indicators allows a combination of centralised and decentralised provision of data as well as work on the data, and it fosters the development of state-of-the-art visualisation and analysis techniques.

---

[4]The IAG is made up of the BIS, the European Central Bank (ECB), Eurostat, the IMF (Chair), the OECD, the United Nations (UN) and the World Bank. "It was established in 2008 to coordinate statistical issues and data gaps highlighted by the global crisis and to strengthen data collection" (Eurostat 2016).

References

Eurostat (2016) Glossary: Inter-Agency Group on economic and financial statistics (IAG). http://ec.europa.eu/eurostat/statistics-explained/index.php/Glossary:Inter-Agency_Group_on_Economic_and_Financial_Statistics_(IAG).

IMF (2011) IMF launches new iPad app to access statistical data. Press Release No. 11/345, 22.09.2011. https://www.imf.org/external/np/sec/pr/2011/pr11345.htm. Accessed 20 Feb 2017

IMF (2014a) A data collection strategy: leveraging SDMX standards. In: Paper presented at the 27th Meeting of the IMF Committee on Balance of Payments Statistics Washington, D.C. 27–29 Oct 2014. https://www.imf.org/external/pubs/ft/bop/2014/pdf/14-08.pdf. Accessed 20 Feb 2017

IMF (2014b) Principal Global Indicators (PGI). IMF/FSB Global Conference on the G-20 Data Gaps Initiative. June 25–26, 2014. .Basel, Switzerland. https://www.imf.org/external/np/seminars/eng/2014/dgi/pdf/m.pdf. Accessed 20 Feb 2017

SDMX (2016) Official homepage of the international SDMX initiative. https://sdmx.org. Accessed 25 Jan 2016

Chapter 10
Conclusion and Outlook

Abstract The volume, complexity and importance of data worlds are increasing explosively. For their profitable use, it is crucial to understand the data, master their rapid growth and put them together to new information structures. However, the information technology (IT) industry has invested little in the development of comprehensive data standards so far.

Statistics as an interdisciplinary science has standards for data management, data documentation, data provision, data protection and, last but not least, the data themselves. These standards are available, distributed worldwide, and they work.

The current data orientation and the rapidly growing data volumes are a great opportunity for statistics to emerge as a central information provider and as a generic discipline for "building knowledge through intelligent evaluation of experience manifested in data". Using this opportunity will be easier with standardisation and SDMX (Statistical Data and Metadata Exchange).

The data universe is rapidly expanding in volume and complexity. Data are increasingly valued, as the well-known quote of Gartner's vice president shows:

Information is the oil of the 21st century, and analytics is the combustion engine.
 (Sondergaard 2011)

Decisions in all areas of life should be evidence based; therefore, a good data foundation to rely on for decision-making and equally good data for the ex post evaluation of the impact of those decisions are needed. High-quality data represent a competitive advantage, and more and more companies are thus becoming *data-driven companies*.

To profitably use the data worlds, it is crucial to master their rapid growth, understand the data, combine it by linking data from different sources and create new information structures that are able to answer all the present and future questions. However, this mastery of the data flood is not a sure-fire success because, unlike other industry branches, the information industry has not taken up the challenge to standardise its most valuable asset: data. A standard for data has not yet been established and global classifications, accessible repositories and a "barcode of information" are lacking. Without further preparation, data worlds are not a modular system in which data sets can easily be assembled to form new products. This is why our vision consists of a well-organised, and therefore well-manageable and well-usable, data world.

Statistics serve as an interdisciplinary auxiliary discipline for almost all areas of science and business and support the idea of "information as a public good". This idea requires standards: standards for collecting data, documenting data, providing data and preserving data confidentiality, and standards for data per se. Statistics offer such standards, which are available, known worldwide, and—and this should be particularly important—have proven their worth in practice. Statistics have successfully and sustainably demonstrated their suitability for daily work, as well as the wide range of possibilities and the international applicability of their concepts. This has led to our belief that the use of these standards should be promoted, in particular the SDMX standard for the realisation of the aforementioned vision of well-organised data worlds.

Statistics should confidently claim these strengths and bring them to the fore. This offensive marketing should consist of the following activities:

- SDMX should be used for the entire value-added chain of information production, i.e. every step of the way: data collection, data quality management, data analysis, data publication, data linking and data documentation.
- New granular data collections—micro data—should, of course, also be classified in SDMX.
- SDMX is very broadly usable, not just for financial and economic data; therefore, the adoption of SDMX by other disciplines is important.
- Public repositories can dramatically speed up distribution.
- It is also helpful to do more marketing in the software industry, since there can be no comprehensive standardisation without their support, be it in the form of the *open source*-approach, by providing their own products, or by providing *software as a service (SaaS)*.
- In addition, there is a need for more (and even more detailed) documentation of the SDMX techniques and their possibilities for the construction, presentation, search and analysis of new information worlds.

The current trend towards data orientation and the rapidly growing data sets provide statisticians with a great opportunity to position themselves as central *information providers*. The authors see statistics as a "generic discipline for building knowledge through intelligent evaluation of experiences manifested by data". Therefore the reflections in this book regarding the creation of comprehensive data worlds do not only apply to economic statistics, but to data in general.

The exploitation of these opportunities will be more successful with the help of standardisation and SDMX.

Reference

Sondergaard P (2011) Gartner says worldwide enterprise IT spending to reach $2.7 trillion in 2012. Analysts discuss key issues facing the IT industry during Gartner symposium/ITxpo 2011, October 16–20, in Orlando. http://www.gartner.com/newsroom/id/1824919. Accessed 20 Feb 2017

Part II
The Statistics Standard SDMX

Why a Second Part?

We deliberately divided this book into two parts because in our experience the detailed explanation of a generic model such as SDMX requires a lot of abstraction as well as technical interest, and might therefore be perceived by some readers as too dry or too difficult to digest. On the other hand, this explanation is essential for the section of our target group that can actively contribute to our vision of the well-arranged data world; thus, it is delivered here in Part II.

As already described in Part I of the book, we consider the ISO standard SDMX to be very well-suited for the construction of a universal, interdisciplinary order system for data. SDMX is more than a technical standard; rather, it is directed at the business professionals, the data experts. It can provide the base for technical implementations on different platforms, different database systems and with different programming languages. In the following chapters we will introduce you to this standard.

Chapter 11
History of SDMX

Abstract The idea underlying the SDMX (Statistical Data and Metadata Exchange) vision dates back to the 1990s and gained its impetus from the preparatory work for the European Monetary Union. The SDMX initiative, founded in 2001, succeeded in establishing the standard in the international statistical data exchange, but also in enhancing it far beyond its original content. The sponsor organisations are constantly working to further develop the standard, promote its use and increase its visibility.

For official statistics, there are great benefits to be gained from expanding the use of SDMX, not only for the data analysts but also for data providers. As a universal data collection format, SDMX could end the era in which each new statistics regulation also defines its own survey formats. With a stronger "industrialisation" of SDMX, additional usage possibilities would also arise outside of statistics.

11.1 The Idea, Its Origin and Its Propagation

The SDMX vision arose on the basis of concepts developed by the Statistical Office of the European Union (Eurostat) in the 1990s. Originally, the initiative ran under the name GESMES (Generic Statistical Message). The basic idea consisted of a generic—that is, a domain-independent, multi-dimensional—data model that should enable Eurostat to standardise the data exchange. Technically this was to be done via a *markup language* used for the self-explanatory structuring of the exchanged data sets. As this development had already taken place before the rise of the Internet and the universal markup language XML, the initiative was based on another markup language popular at that time: EDIFACT.

The initiative achieved rather moderate success in spreading this vision, especially in the software industry. As already mentioned, the understanding of such a generic model requires quite a lot of abstraction. Dissemination mainly through textual (paper-bound) documentation at that time made the new ideas even harder to digest.

Today, it has not really become easier to digest, even though the core body of thought is very simple to understand and is fairly comprehensible for non-mathematicians. But while quite a lot of technical specifications and user guides are to be found on the

© Springer International Publishing AG, part of Springer Nature 2018 73
R. Stahl, P. Staab, *Measuring the Data Universe*,
https://doi.org/10.1007/978-3-319-76989-9_11

Internet, there is still a lack of documentation aimed specifically at non–technical experts, such as easy-to-read papers or even sales brochures winning over potential users. For this reason, this book attempts to provide a descriptive explanation.

The initiative gained real momentum by the end of the 1990s during the preparatory work for the European Monetary Union. In the statistical working groups of the European Monetary Institute (predecessor of the European Central Bank [ECB]), the subset GESMES/CB (the suffix CB stands for central banks) was developed and soon thereafter renamed GESMES/TS (the suffix TS stands for time series). These statistical working groups comprised staff members of the BIS and ECB with the support of Eurostat, external consultants and a number of experts from national central banks, including the Banca d'Italia and the Deutsche Bundesbank. On the basis of this EDIFACT-derived file format (one example can be seen in Fig. 11.1) the entire data exchange in the European System of Central Banks (ESCB) was then established.

This very successful implementation led to the plan mainly initiated by the BIS as a worldwide organisation to establish the GESMES/TS standard as a worldwide ISO standard. The associated vision was to be able to manage the entire global institutional statistical data exchange. The plan then took shape in the form of the SDMX initiative in 2001, when the organisations BIS, ECB, Eurostat, IMF, OECD, UN and the World Bank joined to form a group which is nowadays called the SDMX sponsor organisations. The standard ISO/TS 17369:2005 was established in 2005 and has now matured to ISO 17369:2013. The EDIFACT–GESMES/TS file format is still part of the SDMX standard (under the name SDMX-EDI), and a lot of data are still being exchanged in this format, partly because it is much more compact than the modern XML variants.

SDMX-EDI

The EDIFACT (Electronic Data Interchange for Administration, Commerce and Transport) GESMES/TS (Generic Statistical Message/Time Series) file format (today: SDMX-EDI Is this explanation necessary? The parts are explained in the sentence before.) is a wonderful example of a self-explanatory file format from the times before XML (eXtended Markup Language). In essence, it is a so-called "record format", i.e. the information is transmitted line by line. However, the lines may have different lengths and compositions. For each particular EDIFACT format, a data structure definition (DSD) specifies the types of lines that can exist. Therefore, this also applies for the GESMES/TS format.

For each line, a prefix (the first three characters) defined in the DSD determines the content and the composition. Since transmission formats should be as compact as possible, all textual information is encoded, i.e. replaced by short codes which are also defined in the DSD.

In the case of the specific sample file shown in Fig. 11.1, this means:

(continued)

- Line 1 lists the separators which will structure the information in the following lines.
- The following lines form the message header, which provides information about the sender and content of the file. For example, line 5, which starts with NAD ("name and address") and which names as the sending institution—in a coded form—the Bank for International Settlements (BIS).
- The contents of the transmitted data are written in the then following lines beginning with ARR ("array", meaning array of values) which are always structured according to the pattern "key–observation time–value–value attributes". For example, line 14, which contains the key Q:S:BR:4M:F:I: A:A:TO1:A:AD, the observation time 20134 (fourth quarter of 2013), the value 123, and the value attributes status B, Confidentiality N and "pre-break value" 121. The mystic entry 608 after the observation time specifies the format for the time specification; in this case, a four-digit year followed by the number of the quarter. Even if the format is highly compact, it is easily imaginable that it is self-explanatory after some getting used to by human readers, and is obviously ideal for interpretation by a software.

11.2   The Way to the Global Standard: The SDMX Initiative

In 2001, the SDMX initiative had finally entered the age of the Internet. At the same time, the XML standard had conquered the world of data markup languages. Now it was necessary to maintain the GESMES basic idea of a generic multi-dimensional data model while switching from the traditional data format to the more powerful XML. This also offered the opportunity to drop the specialisation for time series-oriented data, which had dominated the GESMES/TS approach, and to also offer formats for other types of data sets (e.g. cross-sectional data).

Time series-oriented data sets contain data which were observed at periodically repeated times. Each resulting group of related observations is called a time series. Typical time series are weather statistics, which collect daily meteorological parameters, such as temperature or rainfall, for a number of measuring sites. Most of the financial and economic statistics of national central banks are collected on a regular basis and are therefore modelled in time series. Cross-sectional data, on the other hand, are the result of a one-time survey, such as a polling of voters to determine the results on Election Day. However, even those data can, if the surveys are repeated accordingly, lead to time series data sets.

Line	SDMX-EDI message
1	UNA:+.? '
2	UNB+UNOC:3+BR2+5B0+060502:1554+IREF120136++GESMES/TS,
3	UNH+MREF000001+GESMES:2:1:E6'
4	BGM+74'
5	NAD+Z02+BIS'
6	NAD+MR+5B0'
7	NAD+MS+BR2'
8	DSI+BIS_CBS (for template generated output: DSI+IFS_2012_01')
9	STS+3+7'
10	DTM+242:201403131720:203'
11	IDE+5+BIS_CBS'
12	GIS+AR3'
13	GIS+1:::-'
14	ARR++Q:S:BR:4M:F:I:A:A:TO1:A:AD:20134:608:123:B:N:121'
15	ARR++Q:S:BR:4M:F:I:A:A:TO1:A:AR:20134:608:111:B:N:110'
16	ARR++Q:S:BR:4M:F:I:A:A:TO1:A:AT:20134:608:234:B:N:229'
17	ARR++Q:S:BR:4M:F:I:A:A:TO1:A:AU:20134:608:123:B:N:-'
18	ARR++Q:S:BR:4M:F:I:A:A:TO1:A:BE:20134:608:345:A:C'
19	ARR++Q:S:BR:4M:F:I:A:A:TO1:A:CA:20134:608:234:A:N'
...	...

Fig. 11.1 Sample GESMES/TS (Generic Statistical Message/Time Series) file (new name: SDMX-EDI [Statistical Data and Metadata Exchange–Electronic Data Interchange] file) from the *Technical Guidelines for Reporting International Banking Statistics to the BIS* (BIS 2016)

SDMX-ML

The SDMX (Statistical Data and Metadata Exchange) format based on XML (eXtended Markup Language), now called SDMX-ML, is also self-explanatory and thus machine-readable, but is by far not as compact as the previous SDMX-EDI (Electronic Data Interchange) format.

The file shown in Fig. 11.2 contains the same information as the EDIFACT (Electronic Data Interchange for Administration, Commerce and Transport) file in Fig. 11.1. Here, too, we find a message header and information about the sender and the contents of the file. For example, in line 17 the sending institution BIS (Bank for International Settlements).

Lines 23–26 contain the first transmitted value, also naming the key Q:S: BR:4M:F:I:A:A:TO1:A:AD, the observation time 2013–04 (fourth quarter of 2013), the value 123, and the value attributes status B, confidentiality N and "pre-break value" 121.

```
1   ▼<message:StructureSpecificData
2     xmlns:ns="urn:sdmx:org.sdmx.infomodel.datastructure.DataStructure=BIS:BIS_CBS(1.0)ObsLevelDim:TIME_PERIOD"
3     xmlns:structurespec="http://www.sdmx.org/resources/sdmxml/schemas/V2_1/data/structurespecific"
4     xmlns:common="http://www.sdmx.org/resources/sdmxml/schemas/V2_1/common"
5     xmlns:message="http://www.sdmx.org/resources/sdmxml/schemas/V2_1/message" xmlns:xsi="http://www.w3.org/2001/XMLSchema-
6     Instance">
7     ▼<message:Header>
8       <message:ID>IREF120136</message:ID>
9       <message:Test>false</message:Test>
10      <message:Prepared>2014-03-13T17:20:20</message:Prepared>
11      <message:Sender id="BR2"/>
12      <message:Receiver id="5BO"/>
13      ▼<message:Structure structureID="BIS_CBS"
14        namespace="urn:sdmx:org.sdmx.infomodel.datastructure.DataStructure=BIS:BIS_CBS(1.0)ObsLevelDim:TIME_PERIOD"
15        dimensionAtObservation="TIME_PERIOD">
16        ▼<common:Structure>
17          <Ref agencyID="BIS" id="BIS_CBS" version="1.0"/>
18        </common:Structure>
19      </message:Structure>
20      <message:DataSetID>BIS_CBS IFS_2012_01</message:DataSetID>
21    </message:Header>
22    ▼<message:DataSet structurespec:dataScope="DataStructure" xsi:type="ns:DataSetType" structurespec:structureRef="BIS_CBS">
23      ▼<Series FREQ="Q" L_MEASURE="S" L_REF_CTY="BR" CBS_BANK_TYPE="4M" CBS_BASIS="F" L_POSITION="I" L_INSTR="A" REM_MATURITY="A"
24        CURR_TYPE_BOOK="T01" L_CP_SECTOR="A" L_CP_COUNTRY="AD">
25        <Obs TIME_PERIOD="2013-Q4" OBS_VALUE="123" OBS_STATUS="B" OBS_CONF="N" OBS_PRE_BREAK="121"/>
26      </Series>
27      ▼<Series FREQ="Q" L_MEASURE="S" L_REF_CTY="BR" CBS_BANK_TYPE="4M" CBS_BASIS="F" L_POSITION="I" L_INSTR="A" REM_MATURITY="A"
28        CURR_TYPE_BOOK="T01" L_CP_SECTOR="A" L_CP_COUNTRY="AR">
29        <Obs TIME_PERIOD="2013-Q4" OBS_VALUE="111" OBS_STATUS="B" OBS_CONF="N" OBS_PRE_BREAK="110"/>
30      </Series>
31      ▼<Series FREQ="Q" L_MEASURE="S" L_REF_CTY="BR" CBS_BANK_TYPE="4M" CBS_BASIS="F" L_POSITION="I" L_INSTR="A" REM_MATURITY="A"
32        CURR_TYPE_BOOK="T01" L_CP_SECTOR="A" L_CP_COUNTRY="AT">
33        <Obs TIME_PERIOD="2013-Q4" OBS_VALUE="234" OBS_STATUS="B" OBS_CONF="N" OBS_PRE_BREAK="229"/>
34      </Series>
35    </message:DataSet>
36  </message:StructureSpecificData>
```

Fig. 11.2 Sample SDMX-ML (Statistical Data and Metadata Exchange format based on XML [eXtended Markup Language]) file from the *Technical Guidelines for Reporting International Banking Statistics to the BIS* (BIS 2016)

In addition to the technology-driven challenges resulting from the original task of optimally supporting the data exchange between the participating institutions, the SDMX initiative also aimed to generally expand the standard SDMX and to promote its use. Figure 11.3 shows important milestones on the way to the establishment of SDMX as a global data exchange standard of official statistics. It is taken from the current *Roadmap 2020* of the SDMX initiative (SDMX 2016b).

An important landmark in the propagation of SDMX was its approval by ISO as an international standard. In addition, the initiative succeeded in proving the impact of the standard by means of successful pilot projects involving various stakeholders and different thematic areas. In terms of content, the SDMX standard was extended with each new version in such a way that further functionalities important for data sharing were included.

Currently, SDMX is used as a standard for nearly all data exchange in both the European System of Central Banks (consisting of the ECB and associated central banks) and the European Statistical System (consisting of Eurostat and the European statistical offices). SDMX is the carrier format for the IMF's international SDDS Plus initiative , aimed at the worldwide uniform provision of economic and financial data. Important innovations in global accounting systems such as the National Accounts, the Balance of Payments and Foreign Direct Investments are used as an opportunity to classify them according to the rules of the SDMX standard. Thus, SDMX has already come very close to the declared goal recommended by the UN

| 2001 | The seven sponsor organisations BIS, ECB, Eurostat, IMF, OECD, UN and World Bank launch the SDMX initiative

First common statement of the Sponsoring Organisations

Initial Workshop |
| 2002 | First report to UNSC (Common Open standards for the Exchange and Sharing of Socio-economic Data and Metadata: the SDMX Initiative)

Approval of the first work program |
| 2003 | Launch of the first project: Joint External Debt statistics Hub (JEDH) |
| 2004 | Publication of SDMX standard, version 1.0 |
| 2005 | SDMX, version 1.0, accepted by ISO (ISO/TS 17369:2005)

Publication of SDMX standard, version 2.0 |
| 2006 | Go-live of first project (Joint External Debt statistics Hub) |
| 2007 | The sponsor organisations agree on a Memorandum of Understanding (MoU) on the establishment and operation of SDMX

First SDMX Global Conference |
| 2008 | The UN Statistical Commission endorses SDMX as „the preferred standard for exchange and sharing of data and metadata in the global statistical community" |
| 2009 | Second SDMX Global Conference

Publication of first set of „Content Oriented Guidelines" |
| 2011 | Publication of SDMX standard, version 2.1

Third SDMX Global Conference

Founding of the SDMX Statistical Working Group (SWG) and Technical Working Group (TWG) |
| 2013 | SDMX becomes ISO International Standard (IS) 17369

Fourth SDMX Global Conference

Setup of task-force on international data cooperation (TFIDC) by the Inter-Agency Group on Economic and Financial Statistics (IAG)

First release of global Data Structure Definitions for National Accounts, Balance of Payments and Foreign Direct Investment |
| 2014 | Setup of the Ownership Group for SDMX in Macro-Economic Statistics (SDMX-MES OG) |
| 2015 | Extension of the SDMX universe, introducing a new part of the SDMX information model: Validation and Transformation Language 1.0

Launch of the SDMX Global Registry

Revision of SDMX guidelines

Go-live of the international data cooperation pilot on GDP and population

Fifth SDMX Global Conference |

Fig. 11.3 Milestones of the SDMX (Statistical Data and Metadata Exchange) initiative (Annex Key SDMX Milestones, *SDMX Roadmap 2020*, Page 8; SDMX 2016b)

Statistical Commission to become the "preferred standard for exchange and sharing of data and metadata in the global statistical community"[1] (SDMX 2016b).

The sponsor organisations are systematically continuing along this path with the new *SDMX Roadmap 2020*.

11.3   Further Development by the Bodies of the SDMX Initiative

The sponsor organisations have established a committee structure made up of working groups and a secretariat to ensure the prerequisites for continuous further development of the standard. This further development takes place in two branches: the Technical Working Group (TWG; responsible for formal development) and the Statistical Working Group (SWG; responsible for semantic development). Figure 11.4 shows the current body structure of the SDMX initiative. This structure is crucial for the reliable maintenance and steady enhancement of the standard.

The *semantic development* aims to strengthen the standard's power to structure and order data content. The most important assets here are the *Content-Oriented Guidelines*, published for the first time in 2009 (see Fig. 11.4). In particular, the SWG promotes the use of common codelists and the exchange of best practices in the SDMX community.

The *formal development* manifests itself as a continuous technical development, represented by the various versions of the technical standard (see Fig. 11.5). Technical enhancements can be the addition of further file formats (e.g. CSV in addition to EDIFACT and XML), the definition of specifications (e.g. web services, JavaScript Object Notation [JSON]), the implementation of interfaces to standard software (e.g. the statistics software R), conceptual extensions of the SDMX universe (e.g. registries, hierarchical codelists, VTL [Validation and Transformation Language]), or the mapping of the SDMX approach to an adjacent standard (e.g. XBRL [eXtensible Business Reporting Language]). This rapid and multi-faceted further development of SDMX clearly shows that the standard has left the original use case—a data exchange carrier format—far behind.

However, in the world of data processing it is not enough to have established a standard; it is also necessary to ensure that it is used, accepted and even wanted. Especially considering the current scenario of explosively growing data content, the increasing focus on micro data and the rise of the idea of data sharing, it is timely to remember the above-mentioned UN decision of 2008 (see footnote 1). Faced with

---

[1]"The Commission [. . .] recognized and supported SDMX as the preferred standard for the exchange and sharing of data and metadata, requested that the sponsors continue their work on this initiative and encouraged further SDMX implementations by national and international statistical organisations . . ." United Nations Statistical Commission, Report on the Thirty-Ninth Session (26–29 February 2008, section 39/112 b (page 14)).

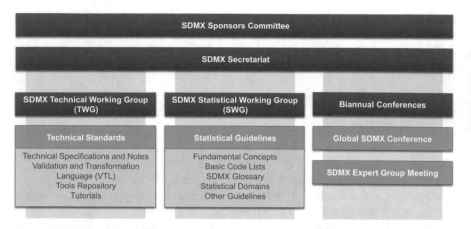

Fig. 11.4 Simplified representation of the current SDMX (Statistical Data and Metadata Exchange) organisation, according to Memorandum of Understanding of the sponsor organisations (original: *SDMX in a Nutshell*, SDMX 2016a)

2004	SDMX Technical Standard, Version 1.0
2005	SDMX Technical Standard, Version 2.0
2009	Content-Oriented Guidelines
2011	SDMX Technical Standard, Version 2.1

Fig. 11.5 Official publications of the SDMX (Statistical Data and Metadata Exchange) initiative (SDMX 2016b)

this challenge, the SDMX sponsor organisations should put their effort into adopting and promoting the use of SDMX, especially for micro data and for cross-institute data sharing.

The SDMX initiative has taken up this mission and has included four development objectives in its *Roadmap 2020* (see Fig. 11.6), which not only further develop the standard in itself but also promote its use and strengthen its publicity. These four main priority areas are aimed at expanding and strengthening the technical foundation, further facilitating data processing with SDMX, fostering the actual use of SDMX for all steps of the statistical business process from data collection to data dissemination, and improving the communication around this standard.

1	Strengthening the implementation of SDMX
2	Making data usage easier via SDMX (especially for policy use)
3	Using SDMX to modernise statistical processes, as well as continuously improving the standards and IT infrastructure
4	Improving communication on SDMX in general and capacity building, including a better interaction between international partners

Fig. 11.6 "Main priority areas" of the *SDMX Roadmap 2020* (SDMX 2016b)

11.4 The Potential in the Use of SDMX as an Information Model

The information model at the core of SDMX shows its power all the more when it is being used outside of the data exchange. This is shown by the fact that more and more sponsor organisations have also already stepped up to use SDMX for their data publication. Several of their large data warehouses available on the Internet are based entirely or partially on SDMX, just a few examples being the Statistical Data Warehouse (SDW) of the ECB, the OECD.stat data platform, the BIS data portal and, last but not least, the statistical time series database on the Deutsche Bundesbank's website. For SDMX-modelled data sets it is comparatively simple to implement visualisations; this is wonderfully demonstrated by the ECB in the provision of statistical data for mobile devices via app (ECBstatsApp, available in app stores).

We have also observed a rapidly increasing use of this standard in software products with a statistical, mathematical or econometric focus, as well as in database solutions specialising in statistical work (e.g. FAME [Forecasting Analysis and Modeling Environment]). This may be due to the fact that some organisations have started to organise their internal data collections based on SDMX. The SDMX multi-dimensional approach provides the perfect starting point for flexible data evaluations, while the standardisation required for the SDMX classification of the topic areas creates the prerequisites for the harmonisation of the data and thus for linkability and integration. This is a clear indication that an SDMX-driven transformation of the statistical landscape is a worthwhile, albeit long-lasting, endeavour.

11.5   The Future: Further Possibilities of Use and Stronger Industrialisation

The benefit of advancing SDMX standardisation is demonstrated in this section, with the help of the statistical business process chain.

In the international statistical community the statistical business process is often represented by the Generic Statistical Business Process Model (GSBPM) (see Sect. 12.8). Simplified, the core of this model consists of six steps, which are fairly similar in each institution responsible for producing statistics (see Fig. 11.7). Step 1 consists of the specification of the requirements. After the conception and design of the statistics (Step 2), the raw data are first collected, i.e. measured or surveyed (Step 3), and, subsequently, the data are processed, which essentially means validating, managing data quality and harmonising (Step 4). Afterwards, results, so-called aggregates—often totals or statistical means—are calculated from the individual data (Step 5). In the final step, the results of the statistics are passed on to the data users or made available to the public (Step 6).

The individual data handled in Steps 3 and 4 are sometimes referred to as micro data or granular data, and the aggregates of Steps 5 and 6 as macro data. Often there is a structural break between the two data worlds, meaning the available micro data follows a different classification system, data model and data format than the required macro data. As a consequence, complex conversions are required. Also different is the way of dealing with the data. For example, micro data are generally subject to strict confidentiality rules; they are only accessible by a restricted user group and exclusively within the boundaries of the pre-agreed purpose of the statistical survey. Generally, micro data did not use to be exchanged or disseminated.

Originally, SDMX started on the "exchange and dissemination" part of the statistics process, so the most powerful implementations are found in the macro data world, for the aggregates (Step 6, also Step 5). When taking into account our description so far, it may have become clear that the information model can equally well be applied to the micro data world (Steps 3 and 4). This would effectively lead to a great advantage for data processing. Finally, the *semantic* differences between the data *models* of the results of a statistical survey and the underlying micro data, which are a common but often unavoidable nuisance, would not be made worse by the *formal* differences of the data *modelling methods*.

This approach is extremely attractive, and not only for the official producers of statistics. A significant improvement could also be made for those obliged to

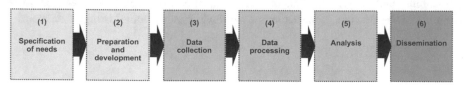

Fig. 11.7 Simplified representation of the Generic Statistical Business Process Model (GSBPM)

periodically deliver data (called "reporting agents" in statistical jargon) if SDMX were set as the base for a new reporting scheme from the design stage. By virtue of its generic principle, SDMX could become a universal format for statistical reporting and end the era in which each new statistical requirement also defined its own data models, specifications and reporting formats. Reporting agents instead could rely on their already existing SDMX-based reporting systems for all new statistics. For this, commercial software products for the requirements of SDMX-based statistical reporting are needed. Accordingly, SDMX would inevitably experience a stronger "industrialisation". Again, with this development, further possibilities of use would also arise beyond the world of official statistics.

References

BIS (2016) Technical guidelines for reporting international banking statistics to the BIS. http://www.bis.org/statistics/bankstatsguide_tech.pdf. Accessed 20 Feb 2017

SDMX (2016a) SDMX in a nutshell. https://sdmx.org/wp-content/uploads/SDMX_map_3_0.jpg. Accessed 25 Jan 2016

SDMX (2016b) SDMX Roadmap 2020. https://sdmx.org/?sdmx_news=sdmx-roadmap-2020. Accessed 25 Jan 2016

United Nations Statistical Commission, Report on the Thirty-Ninth Session (26–29 February 2008). https://unstats.un.org/unsd/statcom/39th-session/documents/statcom-2008-39th-report-E.pdf. Accessed 17 April 2018

Chapter 12
The Main Elements of SDMX

Abstract To explain the information model underlying the SDMX (Statistical Data and Metadata Exchange) standard, its building blocks, the SDMX artefacts, are first introduced. Starting with the basic elements needed to define a data structure, we then gradually add the surrounding elements, such as those used to describe a data set, build a data exchange process, or manage topic areas, actors and processes.

The conceptual world of SDMX is holistic enough to run an SDMX-based data warehouse. Two real-life examples can be found in the European Central Bank and the Deutsche Bundesbank. Of course, SDMX is also suitable for micro data, and with its multi-dimensionality and coded dimensions it provides the ideal interface for any data analysis software.

There are other standards and models apart from SDMX in statistics, some of which we briefly discuss in this chapter: the Generic Statistical Business Process Model (GSBPM), Generic Statistical Information Model (GSIM), Data Documentation Initiative (DDI) and eXtensible Business Reporting Language (XBRL).

In this chapter we take a closer look at SDMX and its body of thought, which up to this point in the book has been presented in a rather sketchy manner. This chapter does not claim to replace a full SDMX documentation, let alone proper training. Rather, we try to familiarise the reader with the design principle underlying SDMX. For the most part, we refer to the currently valid version of the technical specification SMDX 2.1 (SDMX 2013b), which was published in 2011 and consolidated in 2013. However, to facilitate understanding, we have permitted ourselves some simplifications.

The SDMX Information Model is the theoretical foundation of SDMX. To understand this model, we first present its elementary building blocks (referred to as artefacts in SDMX jargon). After that, we follow the line of business use cases, starting from the simplest and moving to the more complex. We begin with the core elements necessary for the design of a data structure, and then gradually add the "surrounding" elements, such as those for the implementation of a data exchange process or for the organisation of whole scientific data areas.

© Springer International Publishing AG, part of Springer Nature 2018   85
R. Stahl, P. Staab, *Measuring the Data Universe*,
https://doi.org/10.1007/978-3-319-76989-9_12

12.1   Elementary Building Blocks

At the beginning of every SDMX classification there is always a set of (mostly) quantitative data to be understood. SDMX provides us with the tools to name, determine and arrange the data points of this set. Take a single data point with the value 3.16. This number by itself is not perceptible to us. The meaning comes with the following description:

Average depth of snow in Austrian ski resorts at altitudes above 2000 m, annual average for the year 2015: 3.16 m (determined by means of observations taken at defined measuring points on the last working day of each month at 8:30 am).

With its information model, SDMX provides a framework into which to fit this description. The most important basic building block is the Concept. And because the Concept is such an important building block of SDMX, we need to go a little deeper into it. And because the Concept is such an important building block of SDMX, we need to go a little deeper into it. The concept could roughly be translated as a "characteristic" or "feature" of a date. Physical characteristics of a person are, for example, their size, weight, blood pressure and cholesterol value. Identifying characteristics are their name, date of birth, place of birth and social security number. The properties of a mortgage loan are, for example, the interest rate, redemption plan, maturity and registered mortgage. A concept is thus a term for everything that is elementary or at least helpful to the understanding of a specific datum or data set. For each property that describes the data point in more detail, an SDMX Concept acts as—technically speaking—the container. In the above-mentioned sample data point, there is one property, which is "Austrian"; the corresponding container term— the concept—would be "country". The qualitative description in the brackets ("determined by means of. . .") also describes the datum more precisely. The matching concept could be called the "measuring method".

Concepts differ greatly in content, shape and possible representations. The concept "measuring method" allows for free-text data and is therefore referred to as an *uncoded Concept* in SDMX jargon. On the other hand, for the concept "country", only values which belong to a predefined list of permitted entries are allowed. Such a concept is called a *coded Concept*, and the list of permitted entries is a *Codelist*. The codes of such a codelist consist, for practical reasons, of a short, unique ID and a longer textual description—or several descriptions in multi-lingual scenarios. For example, code ID = AT, code description (English) = Austria, code description (German) = Österreich and so on. (Actually, this code example was taken from the official list of ISO countries, which is not only ideally suited for use but in reality often used in SDMX DSDs. The question of how difficult it is to agree on the concrete entries in such an international multi-purpose country codelist is quite complex and would require a separate book.)

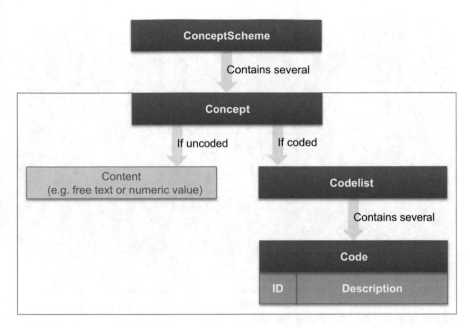

Fig. 12.1 ConceptScheme and Concept

In addition to the coded and free-text variants, there are also various other forms of representation, for example numeric values, dates or texts with a given string pattern (a string pattern is a format specification such as "two letters, followed by six digits, followed by one of the special characters * or #").

In order to handle the management of related concepts, these are often maintained in a common package—a ConceptScheme (see Fig. 12.1). This method for managing objects of the same type—putting them in a [. . .]Scheme—can be found in many places in the SDMX Information Model.

12.2 Defining a Data Structure Definition

The basic building block Concept is now being used extensively to structure an existing data set or—in SDMX jargon—to define its DSD. There are various roles in a DSD which need to be filled:

- *Measure*: the actual measured value, the content of the data points (*observations*)
- *Dimension*: the classifying or uniquely identifying properties of the data points of a data set.
- *DataAttribute*: additional properties that describe the data points in more detail.

Type of aggregation ("Average")	Dimension
Observation ("Depth of snow ... in ski resorts")	Measure
Country ("Austria")	Dimension
Altitude ("above 2,000 m")	Dimension
Frequency ("Annual")	Dimension
Time (or date) of observation ("2015")	(Time) Dimension
Unit (Metre)*	DataAttribute
Measuring method ("measurements at predefined locations, at 8:30 in the morning...")	DataAttribute

*Theoretically, the unit (metre) could also be understood as a dimension. An argument for this choice would be its necessity - without specifying the unit, the numerical value can not be interpreted correctly. However, the reason for the classification as an attribute is the circumstance that the observed measure per se does not change by changing the unit alone (for example, a conversion to feet or centimetres).

Fig. 12.2 Dimensions and attributes for the data point "Average depth of snow in Austrian ski resorts above 2000 m, in 2015: 3.16 m (measuring method:...)"

If we deconstruct the earlier example according to this scheme, we obtain the elements shown in Fig. 12.2.

The dimensions are formed from coded concepts because they form the axes of the coordinate systems of the data sets based on this DSD. Each data point (observation) in this coordinate system is unambiguously defined by the specification of the codes for all—in the above sample case, five—dimensions. If a valid code is given for each dimension, a unique identifier—the key—emerges for this data point (see Fig. 12.3).

SDMX has had quite a few extensions since version 1.0 of the standard, many with the aim of gaining more flexibility in the modelling of data sets, or of being able to describe unusually shaped data sets. In the latest version 2.1, a DSD can contain more than one measure. To make this possible, a MeasureDimension is used, which in turn contains a list of measured variables. When there is only one measured value, which is the case most of the time, it is called the PrimaryMeasure.

A DataAttribute can refer to different levels of the DSD, such as the entire data set, a certain subset (group) or even a single data point (observation). The exact level of attachment is defined in an AttributeRelationship (see Fig. 12.4).

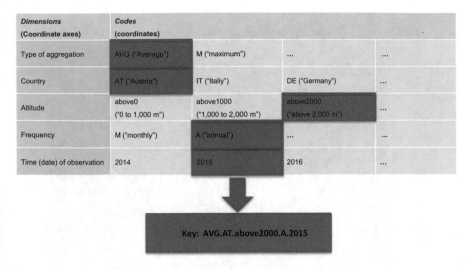

Dimensions (Coordinate axes)	Codes (coordinates)			
Type of aggregation	AVG ("Average")	M ("maximum")	...	...
Country	AT ("Austria")	IT ("Italiy")	DE ("Germany")	...
Altitude	above0 ("0 to 1,000 m")	above1000 ("1,000 to 2,000 m")	above2000 ("above 2,000 m")	...
Frequency	M ("monthly")	A ("annual")	...	...
Time (date) of observation	2014	2015	2016	...

Key: AVG.AT.above2000.A.2015

Fig. 12.3 Coordinate axes, coordinates and key

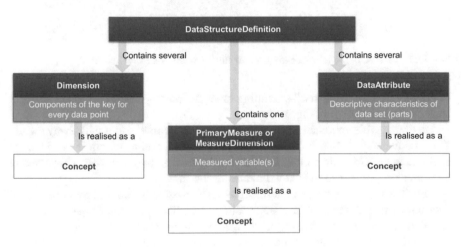

Fig. 12.4 DataStructureDefinition

12.3   Creating a Data Set According to a Data Structure Definition

What is to be done to create a data set that corresponds to a DSD such as the one described in Sect. 12.2? The simplest way would be to first label the DataSet element, preferably with a data set identifier (DSI), and then list all existing individual data points directly under it, in the form of key–measure pairs; or, in SDMX jargon, the KeyValue–ObservationValue. The key is, as already described, the combination of the respective codes of all dimensions. If we applied this to our

DataSet		
Observation		
	KeyValue	AVG:AT:above2000:A:2015 or, in words,
		Type of aggregation = average, country = Austria, altitude = above 2,000 m, frequency = annual, time = 2015
	ObservationValue	3.16
Observation		
	KeyValue	AVG:AT:above2000:A:2015 or, in words,
		Type of aggregation = average, country = Austria, altitude = above 2,000 m, frequency = annual, time = 2016
	ObservationValue	2.12
More Observations …		

Fig. 12.5 DataSet, observations ungrouped, for the "Certainty of Snow Statistics"

"Certainty of Snow Statistics"—omitting other attributes—we would receive a result such as the one shown in Fig. 12.5.

Obviously, this approach has its advantages. It is simple, and it will always work. The disadvantage is that it enforces many repetitions when writing these data, for example in an XML file. In the above example, the first two data points have nearly the same key, except for one key dimension: the time.

This is why, in practice, an intermediate level is often introduced by grouping all data points that differ only in a single dimension. This particular type of group is called a Series. If the dimension chosen is the time dimension, it is called a TimeSeries—as it is truly a time series. A series is described by the specification of its key, which obviously lacks only one dimension compared to the observation. To specify the observations within the series only the code of the missing dimension is needed—they can be listed as pairs of (Time)KeyValue–ObservationValue. Figure 12.6 shows what this means for our example of "Certainty of Snow Statistics".

To complete the model, you now only have to add the DataAttributes previously defined in the DSD. As in the DSD, the attributes can be attached at practically every level. There is quite often the situation that a certain attribute applies to a whole group of observations or series; in our example this could mean an attribute for all data points relating to the country Austria. In order to create an anchor point for such attributes, the idea of a group is introduced. A group is a subset of the data set that is defined by fixed codes for some (meaning several, but not all) dimensions—the combination of those fixed codes forms the GroupKey. Such groups can appear

DataSet		
Time Series		
	SeriesKey	AVG:AT:above2000:A or, in words,
		Type of aggregation = average, country = Austria, altitude = above 2,000 m, frequency = annual (leaving out the time dimension)
	Observation	
		TimeKeyValue	2015
		ObservationValue	3.16
	Observation	
		TimeKeyValue	2016
		ObservationValue	2.12
	More Observations …	
More Series …		

Fig. 12.6 DataSet, observations grouped in time series, for the "Certainty of Snow Statistics"

parallel to the series in the data set; usually they are just used as attachment points for attributes. An overview of the SDMX elements of a DataSet is shown in Fig. 12.7.

12.4   Data Sets Are Exchanged Between Parties

When introducing the SDMX elements that are necessary to prepare the data sets defined in Sect. 12.3 for a transmission from a sender to a recipient, it becomes clear that SDMX was initially designed to advance the professional data exchange of multiple interconnected institutions. In addition to the SDMX artefacts that describe the data structure and encapsulate the data contents, SDMX has a number of additional elements to administer a decentralised managed multi-party data exchange. In this section we provide just a few examples.

We already learned that a DataSet is built according to a DataStructureDefinition. The institution that provides the data—the DataProvider—agrees (usually with the recipients) on a ProvisionAgreement, which specifies which kind of data they provide at which times. Regular data transfers are recorded in a DataflowDefinition, which describes not only the structure (i.e. DataStructureDefinition) of the DataSets to be transferred, but also additional information, such as a more detailed description of the content or so-called constraints, i.e. conditions for the data of the data sets.

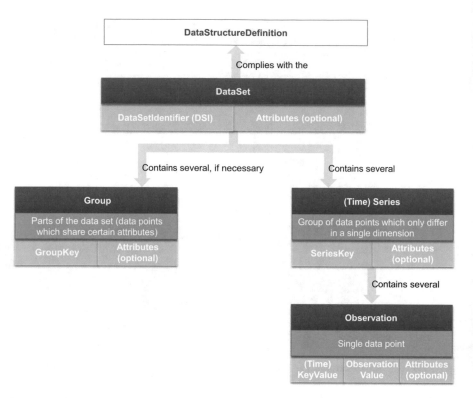

Fig. 12.7 DataSet

These artefacts can be used to describe the real scenarios of a multi-party data exchange, such as that of international official statistics. For example, it is possible for Eurostat to define a DSD to be used by all parties for a set of economic indicators. The European national statistical offices (DataProviders) then pledge to transmit the respective national figures for these indicators (DataSets). This would be modelled by a ProvisionAgreement and an associated DataflowDefinition, with the constraint that, for example, the German Federal Statistical Office only transmits the German figures, the French equivalent INSEE (Institut national de la statistique et des études économiques) only the French numbers and so on (see Fig. 12.8).

Usually in such an arrangement, the DSDs, provision agreements and data flow definitions are transferred only once at the beginning or each time the process has to be adapted. Afterwards only data sets are transmitted.

For the actual sending of data sets, a number of possible file formats are available, the most powerful of which is the XML-based format called SDMX-ML. But other formats have been, and are still being, used in the SDMX world as well, such as the older EDIFACT record format or domain-specific CSV files. SDMX-ML carries all the advantages of an XML derivative format, namely that XML validation via XML schema files can be used to determine the formal correctness of a file with regard to the SDMX framework or even the DSD.

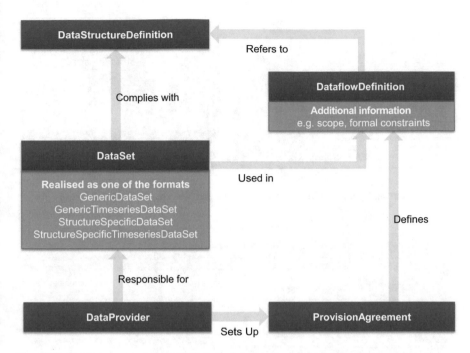

Fig. 12.8 DataSet, surrounded by artefacts designed to model data exchange processes

The formal check for "SDMX correctness" for a given data file can be achieved in two variants: one is the *generic* check—the file is checked to see if the SDMX artefacts such as dimension, key and observation are used correctly. There are general schema files that validate SDMX artefacts and their use. The other one is the *structure-specific* check: the data file is checked to see whether its structure corresponds to a given DataStructureDefinition. For this purpose, schema files that are designed specifically for a DSD are to be used, which translate the concrete specifications of the DSD (such as the number and name of the dimensions, the codelists, etc.) into XML schema elements. Obviously, the second variant is much more enlightening as regards the correctness of the data file. While the generic schema files are created and maintained by the official bodies of the SDMX initiative, structure-specific schemas must be provided by the institution that introduces a new DataStructureDefinition and plans to use it.

The two variants result in two different message types: the generic as well as the structural-specific message type. And since SDMX has traditionally been used extensively for the exchange of time series data, there is also a variant optimised for time series for each of the two message types (see Figs. 12.9 and 12.10).

The "M" in SDMX stands for "metadata"; thus, there are not only file formats for the exchange of data sets but also other message types for exchanging the metadata, such as the DSD, which must be sent at least once, namely when the actual data exchange is established.

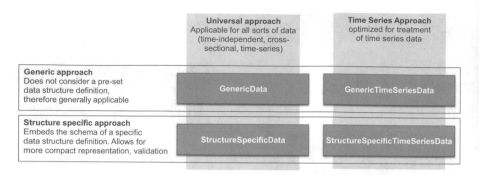

Fig. 12.9 Message types for SDMX (Statistical Data and Metadata Exchange) data sets

```
▼<message:StructureSpecificTimeSeriesData xmlns:exr="urn:sdmx:org.sdmx.infomodel.datastructure.DataStructure=ECB:ECB_EXR_NG(1.0):ObsLevelDim:TIME_PERIOD"
   xmlns:message="http://www.sdmx.org/resources/sdmxml/schemas/v2_1/message"
   xmlns:data="http://www.sdmx.org/resources/sdmxml/schemas/v2_1/data/structurespecific" xmlns:common="http://www.sdmx.org/resources/sdmxml/schemas/v2_1/common"
   xmlns:xsi="http://www.w3.org/2001/XMLSchema-instance" xsi:schemaLocation="http://www.sdmx.org/resources/sdmxml/schemas/v2_1/message
   ../../../../schemas/SDMXMessage.xsd urn:sdmx:org.sdmx.infomodel.datastructure.DataStructure=ECB:ECB_EXR_NG(1.0):ObsLevelDim:TIME_PERIOD ecb_exr_ng_ts.xsd">
   ▼<message:Header>
       <message:ID>Generic</message:ID>
       <message:Test>false</message:Test>
       <message:Prepared>2010-01-04T16:21:49+01:00</message:Prepared>
       <message:Sender id="ECB"/>
       ▼<message:Structure structureID="STR1" dimensionAtObservation="TIME_PERIOD"
           namespace="urn:sdmx:org.sdmx.infomodel.datastructure.DataStructure=ECB:ECB_EXR_NG(1.0):ObsLevelDim:TIME_PERIOD">
           ▼<common:Structure>
               <Ref agencyID="ECB" id="ECB_EXR_NG" version="1.0"/>
           </common:Structure>
       </message:Structure>
   </message:Header>
   ▼<message:DataSet data:structureRef="STR1" xsi:type="exr:DataSetType" data:dataScope="DataStructure">
       ▼<Series FREQ="M" CURRENCY="CHF" CURRENCY_DENOM="EUR" EXR_TYPE="SP00" EXR_VAR="E" DECIMALS="4" UNIT_MEASURE="CHF" UNIT_MULT="0" COLL_METHOD="Average of
           observations through period" TITLE="ECB reference exchange rate, Swiss franc/Euro">
           <Obs TIME_PERIOD="2010-08" OBS_VALUE="1.3413" OBS_STATUS="A" CONF_STATUS_OBS="F"/>
           <Obs TIME_PERIOD="2010-09" OBS_VALUE="1.3089" OBS_STATUS="A" CONF_STATUS_OBS="F"/>
           <Obs TIME_PERIOD="2010-10" OBS_VALUE="1.3452" OBS_STATUS="A" CONF_STATUS_OBS="F"/>
       </Series>
       ▼<Series FREQ="M" CURRENCY="GBP" CURRENCY_DENOM="EUR" EXR_TYPE="SP00" EXR_VAR="E" DECIMALS="5" UNIT_MEASURE="GBP" UNIT_MULT="0" COLL_METHOD="Average of
           observations through period" TITLE="ECB reference exchange rate, U.K. Pound sterling /Euro">
           <Obs TIME_PERIOD="2010-08" OBS_VALUE="0.82363" OBS_STATUS="A" CONF_STATUS_OBS="F"/>
           <Obs TIME_PERIOD="2010-09" OBS_VALUE="0.83987" OBS_STATUS="A" CONF_STATUS_OBS="F"/>
           <Obs TIME_PERIOD="2010-10" OBS_VALUE="0.87637" OBS_STATUS="A" CONF_STATUS_OBS="F"/>
       </Series>
       ▼<Series FREQ="M" CURRENCY="JPY" CURRENCY_DENOM="EUR" EXR_TYPE="SP00" EXR_VAR="E" DECIMALS="2" UNIT_MEASURE="JPY" UNIT_MULT="0" COLL_METHOD="Average of
           observations through period" TITLE="ECB reference exchange rate, Japanese yen/Euro">
           <Obs TIME_PERIOD="2010-08" OBS_VALUE="110.04" OBS_STATUS="A" CONF_STATUS_OBS="F"/>
           <Obs TIME_PERIOD="2010-09" OBS_VALUE="110.26" OBS_STATUS="A" CONF_STATUS_OBS="F"/>
           <Obs TIME_PERIOD="2010-10" OBS_VALUE="113.67" OBS_STATUS="A" CONF_STATUS_OBS="F"/>
       </Series>
       ▼<Series FREQ="M" CURRENCY="USD" CURRENCY_DENOM="EUR" EXR_TYPE="SP00" EXR_VAR="E" DECIMALS="4" UNIT_MEASURE="USD" UNIT_MULT="0" COLL_METHOD="Average of
           observations through period" TITLE="ECB reference exchange rate, U.S. dollar/Euro">
           <Obs TIME_PERIOD="2010-08" OBS_VALUE="1.2894" OBS_STATUS="A" CONF_STATUS_OBS="F"/>
           <Obs TIME_PERIOD="2010-09" OBS_VALUE="1.3067" OBS_STATUS="A" CONF_STATUS_OBS="F"/>
           <Obs TIME_PERIOD="2010-10" OBS_VALUE="1.3898" OBS_STATUS="A" CONF_STATUS_OBS="F"/>
       </Series>
   </message:DataSet>
</message:StructureSpecificTimeSeriesData>
```

Fig. 12.10 SDMX-ML (Statistical Data and Metadata Exchange format based on XML [eXtended Markup Language]) file, structure-specific scheme, optimised for time series data (SDMX 2013a)

12.5 The Greater Perspective: Management of Information, Topic Areas, Stakeholders and Processes

At this point our more detailed look at the SDMX specification in version 2.1 ends, but not without the closing remark that we have only introduced a small part of the SDMX universe. The body of thought inherent in SDMX has grown and matured over more than ten years and therefore covers much larger areas, which we can only hint at in the following paragraphs. A better idea of the vastness of the SDMX Information Model, can be found on various websites, and in particular on an interactive website called Clickable SDMX, which is made available by the UN Economic Commission for Europe (UNECE) (UNECE 2016).

In the following subsections, we provide a few examples of the areas not explained in detail here.

Facilitating Data Searches via Central Information Portals
The SDMX model offers the possibility to store the DSDs, together with information on data providers and data flows, in a so-called registry. A registry is a central entry point to all information available for a certain area—it is often referred to as an information hub. The registry usually contains only metadata, i.e. information defining the data, and references to the decentrally stored data sets. Data users can use the registry to search for the desired information; some even offer subscription or notification services. The current DSDs can be downloaded from the registry and—via a link to websites or web services of the actual data sources—also the corresponding data sets.

Documenting Data Even Better Using Additional Metadata
In addition to the so-called "structural metadata" (basically the data structure definition), other information about a data set might also be worth sharing, such as information about methodology, data quality, source and contact details. Such data is called "referential metadata" in the SDMX jargon, and it is usually given in free-text format. SDMX also provides a framework for these metadata, which consists, among other things, of a Metadata Structure Definition (MSD) and appropriate additional concepts.

An example of this is the Euro SDMX Metadata Structure (ESMS), developed by Eurostat (see Fig. 12.11).

Organise Topics in Thematic Areas
SDMX has several artefacts for the organisation of topics (subject-matter domains). As regards reporting, data flows can be inserted into a so-called reporting taxonomy, which allows for a context-driven hierarchical structure of the data reported.

Managing Stakeholders of Data Exchange and Data Sharing
Since SDMX was originally designed for data exchange between institutions, there are also artefacts for the management of the parties in these data exchange processes: there are organisation schemes that may include data consumers and data providers. The actors who have taken on the development and administration of subject-specific SDMX artefacts, such as data structure definitions and codelists, are called agencies in the SDMX universe.

Further Enhancing Flexibility in the Modelling of Data
The SDMX vocabulary is continually being extended; the latest additions include formula languages for validations and calculations, as well as tools for the modelling of business processes. A typical example of the continuous development is the codelist, which initially did not have any substructures. It soon became clear, however, that a hierarchy, i.e. the formation of groups and subgroups of codes, is required. The next extension of the codelist contained a simple hierarchy, which

No	Concept name	Concept code	Contains the following concepts
1	Contact	CONTACT	Contact organisation, Contact organisation unit, Contact name, Contact person function, Contact mail address, Contact email address, Contact phone number, Contact fax number
2	Metadata update	META_UPDATE	Metadata last certified, Metadata last posted, Metadata last update
3	Statistical presentation	STAT_PRES	Data description, Classification system, Sector coverage, Statistical concepts and definitions, Statistical unit, Statistical population, Reference area, Time coverage, Base period
4	Unit of measure	UNIT_MEASURE	
5	Reference period	REF_PERIOD	
6	Institutional mandate	INST_MANDATE	Legal acts and other agreements, Data sharing
7	Confidentiality	CONF	Confidentiality – policy, Confidentiality - data treatment
8	Release policy	REL_POLICY	Release calendar, Release calendar access, User access
9	Frequency of dissemination	FREQ_DISS	
10	Accessibility and clarity	ACCESSIBILITY_CLARITY	News release, Publications, On-line database, Micro-data access, Other, Documentation on methodology, Quality documentation
11	Quality management	QUALITY_MGMNT	Quality assurance, Quality assessment
12	Relevance	RELEVANCE	User needs, User satisfaction, Completeness
13	Accuracy and reliability	ACCURACY	Overall accuracy, Sampling error, Non-sampling error
14	Timeliness and punctuality	TIMELINESS_PUNCT	Timeliness, Punctuality
15	Coherence and comparability	COHER_COMPAR	Comparability – geographical, Comparability - over time, Coherence - cross domain, Coherence – internal
16	Cost and burden	COST_BURDEN	
17	Data revision	DATA_REV	Data revision – policy, Data revision – practice
18	Statistical processing	STAT_PROCESS	Source data, Frequency of data collection, Data collection, Data validation, Data compilation, Adjustment
19	Comment	COMMENT_DSET	

Fig. 12.11 Euro SDMX Metadata Structure 2.0, developed By Eurostat (modified, shortened) (Eurostat 2016a)

allowed for a maximum of one parent code for each code. In the newest version of the standard, there is the very flexible instrument of the hierarchical codelist, which allows significantly more variations. Similarly, the concepts of multi-lingualism or versioning of artefacts were gradually added to the standard.

12.6   The SDMX-Based Data Warehouse

The SMDX standard's body of thought is powerful enough to serve as the foundation for the full-blown implementation of an SDMX-based data collection—the SDMX-based data warehouse. Two current examples are found in the European Central Bank and the Deutsche Bundesbank.

In both cases, SDMX's key concepts for the design and maintenance of DSDs had to be translated into a relational database schema, thus creating data storage that can contain SDMX-classified data and make them searchable by using their dimensions. Both data warehouses enable flexible navigation and filtering of data according to their SDMX structure via online portal or web services. They include user interfaces as well as programming interfaces for the maintenance of the DSDs and thus offer the opportunity to advance the semantic harmonisation of the data (e.g. by adapting and harmonising the codelists), which is already made easier simply on account of storing the data as SDMX data.

Consequently, the next step should be the linking of SDMX data sets along their natural seam lines created by their SDMX classifications, i.e. the codelists they share. This rather mechanical application is quite easy; however, it is important to note what has already been described in Sect. 2.3: a purely technical link between two data sets may be entirely feasible but completely nonsensical. To the naked eye, the data sets are perfectly matched, but the final assessment can only be made on the basis of an understanding of the data contents. Therefore, content-related information is a central component of an integrated data collection; this is where one reaches the borderline of formally manageable, IT-accessible information, and enters the field of intelligent data analysis. Here, added value is not possible without knowledge of the data content, collection methods, details, special features on so on.

12.7 Applicability of SDMX for Micro Data

Experts in the field of data production and processing often differentiate between various types of data: operational data, process data, analytical data, research data, macro data, micro data, supervisory data, statistical data, master data, metadata and so on.

In the context of standardisation, we do not consider this distinction to be meaningful. Each type of data ultimately consists, as already described, of identifying characteristics (key dimensions), additional descriptive characteristics (attributes) and observation values of the phenomenon itself. The usage of the data may be different, but this does not require different structuring, documentation or modelling of the data. Therefore, there is a very simple answer to the question of whether SDMX is also suitable for micro data: of course it is!

In the case of micro data, also called granular data, there are of course key dimensions that identify individual entities, such as the ID for a single security (ISIN), single person (social security number) or single vehicle (licence plate number). In contrast, for macro data, the key dimensions identify certain groups or categories, such as industry branches, sectors of the economy or groups of people.

When dealing with micro data, it is to be expected that the finer granularity usually leads to much larger amounts of data. For the data exchange processes, lengthy XML data formats are then often regarded as too "chatty", compared to very compact formats such as CSV. To counter this phenomenon, the *SDMX Roadmap 2020* foresees the promotion of easy-to-use SDMX-compatible file formats such as CSV. The most important thing about these formats is that, despite their compact size, the data structure defined by the SDMX metadata is still complied with. Thus, this would obviously not be a "free" kind of CSV, but a well-defined CSV matching the SDMX DSD.

Large quantities of data should be addressed with suitable software, for example the BI, data warehouse or Big Data tools, which have already been mentioned several times in this book. With its multi-dimensional nature and its coded key dimensions, SDMX provides the ideal groundwork for BI tools, for the design of BI cubes and associated loading processes, the so-called ETL [Extract–Transform– Load] processes. Based on this, a generic process for transferring an SDMX data set to a BI cube can easily be created, and since SDMX is by nature generic, this process is not only applicable to this data set, but in general!

12.8   SDMX and Neighbouring Standards

In Sect. 5.1, we described the difficulties that hinder the rise of any potential standard. One of those is the fact that a newly proposed standard is usually not the only candidate in its field. This is no different in statistics. There, also various concepts and models have become more or less widespread. We provide a summary of some of them here.

Generic Statistical Business Process Model (GSBPM)
Official statistics and the corresponding, historically grown data world are far from being fully SDMX-classified. They know other standards or models, some more closely aligned to the processes than to the data.

The Generic Statistical Business Process Model (GSBPM) was developed as part of the Common Metadata Framework (CMF) by the European Statistician Steering Group on Statistical Metadata (see www.unece.org/stats/cmf). The GSBPM supplies a reference model that contains all business processes required for the production of official statistics. The uniform usage of the terms defined in this process model facilitates the communication between different institutions. The formal classification of all work steps necessary for statistics production is the ideal starting point for considerations regarding technical and methodological standardisation, process automation, quality control and synergies (see Fig. 12.12).

Generic Statistical Information Model (GSIM)
The Generic Statistical Information Model (GSIM) was created as a container for all the information objects and data flows of the GSBPM. As an information model, it is also devoted to the definitions, relations and attributes of the data worlds and tries to bridge the gap between neighbouring standards. "GSIM aligns with relevant standards such as DDI and SDMX" (UNECE 2017).

Data Documentation Initiative (DDI)
From the domain of the social sciences comes a fairly universal data documentation standard, the primary focus of which is on the scientific exploitation of data sets.

The Data Documentation Initiative (DDI) is a metadata standard originally aimed at comprehensively documenting data sets for users—especially in the field of statistical surveys. Since its introduction, the DDI approach has expanded to reflect the entire life cycle of a data set. The standard is developed and maintained by the

Quality Management / Metadata Management							
Specify Needs	**Design**	**Build**	**Collect**	**Process**	**Analyse**	**Disseminate**	**Evaluate**
Identify Needs	Design outputs	Build collection instrument	Create frame and select sample	Integrate data	Prepare draft outputs	Update output systems	Gather evaluation inputs
Consult and confirm needs	Design variable descriptions	Build or enhance process components	Set up collection	Classify and code	Validate outputs	Produce dissemination products	Conduct evaluation
Establish output objectives	Design collection	Build or enhance dissemination components	Run collection	Review and validate	Interpret and explain outputs	Manage release of dissemination products	Agree on action plan
Identify concepts	Design frame and sample	Configure workflows	Finalise collection	Edit and impute	Apply disclosure control	Promote dissemination products	
Check data availability	Design processing and analysis	Test production system		Derive new variables and units	Finalise outputs	Manage user support	
Prepare business case	Design production systems and workflow	Test statistical business process		Calculate weights			
		Finalise production system		Calculate aggregates			
				Finalise data files			

Fig. 12.12 Schematic representation of the Generic Statistical Business Process Model (GSBPM), version 5.0, published 2013 (UNECE 2013)

international Data Documentation Initiative Alliance, a consortium whose members are composed of universities, research institutes, data archives and statistical organisations. The DDI standard is mainly used in social science research. Here, too, there is keen awareness of the neighbouring standards and a wish for good cooperation. The compatibility of DDI and SDMX has been intensively investigated and improved by the initiatives of both sides.

eXtensible Business Reporting Language (XBRL)

In order to handle balance sheet reporting, a standard that has been strongly propagated by the industry has been established.

XBRL basically aims to facilitate financial reporting. The standard therefore focuses primarily on the beginning of the value-added chain of (statistical) data processing and analysis—the data collection. XBRL supports the preparation of the balance sheet report of a single company, its transfer and its reception by a data-collecting institution.

In order to support these use cases as comprehensibly as possible, the XBRL organisation (XBRL International, headquartered in the USA) adopted a very ambitious approach, which goes way beyond purely formal and semantic data descriptions. The DSDs—called taxonomies in XBRL jargon—cover additional tasks such as the validation and presentation layers. In most data collection realisations known to us, these additional features are hardly used. Nevertheless, their inclusion makes working with the taxonomies a somewhat very demanding, extremely complex craft.

The Relationship with SDMX

In Sect. 5.1, we noted that new standards usually do not establish themselves in uncharted territory, and most often a potential standard is not the only or first candidate in its field. It is therefore not surprising that all standard initiatives in the area of statistics, despite their willingness to cooperate, still try to stake their claim.

To date, SDMX has made its major contributions at the end of the statistical business process, in the steps dealing with data publication, data exchange and data dissemination, and therefore has mostly handled the results, i.e. the aggregates—macro data. So it is no surprise that the standards originally resident in the field of micro data suggest that SDMX is only suitable for aggregate data: "... where DDI is aimed at solving problems with the documentation of research, and across the micro-data lifecycle, SDMX is concerned with creating efficiencies around the exchange of aggregate data" (Gregory and Heus 2007). On the other hand, the XBRL community tries to set itself apart from the SDMX approach by referring to the more comprehensive approach of the XBRL exchange format described here.

We believe that we have adequately demonstrated in this book that SDMX is a valid, high-potential standard also suitable for the micro data world, and that the SDMX approach is a holistic foundation for the organisation of comprehensive data worlds and for the support of the entire value-added chain, not just for a specific part of the business process.

References

Eurostat (2016) Euro-SDMX metadata structure ESMS. http://ec.europa.eu/eurostat/data/metadata/
    metadata-structure. Accessed 20 Feb 2017
Gregory A, Heus P (2007) DDI and SDMX: complementary, not competing standards. Version 1.0.
    Jul 2007. http://www.opendatafoundation.org/papers/DDI_and_SDMX.pdf. Accessed 20 Feb
    2017
SDMX (2013a) SDMX 2.1 Technical specifications—consolidated version 2013—Section 3B—
    SDMX-ML. XML schemas, samples, WADL and WSDL (update 2013). http://sdmx.org/wp-
    content/uploads/SDMX_2-1-1_SECTION_3B_SDMX_ML_Schemas_Samples_201308.zip.
    Accessed 25 Jan 2016
SDMX (2013b) SDMX 2.1 Technical specifications—consolidated version 2013. https://sdmx.
    org/?page_id=5008. Accessed 25 Jan 2016
UNECE (2013) Statistical metadata (METIS)/METIS-wiki/generic statistical information model/
    the Generic Statistical Business Process Model. Created, and most recently changed, by Thérèse
    Lalor on 23 December 2013. http://www1.unece.org/stat/platform/display/metis/The+Generic
    +Statistical+Business+Process+Model. Accessed 20 Feb 2017
UNECE (2016) Clickable SDMX/Clickable SDMX Home/SDMX Information Model. Created by
    Chris Jones. Most recently changed by Laura Vignola on 8 July, 2016. http://www1.unece.org/
    stat/platform/display/ClickSDMX/Clickable+SDMX+Home. Accessed 20 Feb 2017
UNECE (2017) The Generic Statistical Information Model (GSIM). http://www1.unece.org/stat/
    platform/download/attachments/75564118/First%20GSIM%20Brochure%201_1.pdf?api=v2.
    Accessed 20 Feb 2017.

Chapter 13
Working with SDMX

Abstract Although SDMX (Statistical Data and Metadata Exchange) has spread significantly throughout the world of official statistics, most of the software products based on it have been self-developed by statistical institutions, and are therefore proprietary. However, there is a fair amount of open-source software provided by the SDMX community free of charge, which can be very helpful to those approaching the subject.

These tools have a very different degree of maturity and often offer overlapping functionalities. But with the help of this software collection, the following requirements can be met: create new and manage existing data structures; file and manage SDMX datasets; edit SDMX files; work with the data in SDMX datasets; and store information about SDMX datasets centrally. On top of that, programming libraries can also be found to use for in-house development.

The most beautiful models remain only theory, if there are no implementations, because the ideas can unfold their power and their utility only in real products. But how can an interested party start working with SDMX and get to know the model in practical terms?

Although SDMX has achieved a respectable distribution in the world of statistics, the corresponding products have mostly been developed in specialised units of statistical entities, with the result that almost all of them are proprietary.

However, especially for newcomers searching to approach the topic, open-source components are important, and the SDMX community provides several free of charge. The official SDMX website, for example, lists a whole range of software products developed by SDMX-using institutions on its web page "Software Tools for SDMX Implementers and Developers" (SDMX 2018). These tools have a varying degree of maturity and often offer an overlapping range of functionalities. We attempt to classify them here, according to the listing at April 2018.

© Springer International Publishing AG, part of Springer Nature 2018
R. Stahl, P. Staab, *Measuring the Data Universe*,
https://doi.org/10.1007/978-3-319-76989-9_13

Creating new data structures, managing existing data structures

Data Structure Wizard (DSW)	Eurostat	Wizard for creating your own SDMX data structure; helps you to create, edit and test SDMX artefacts
XSD Generator	Eurostat	Generates the XML schema corresponding to a given SDMX data structure definition
Mapping Assistant	Eurostat	Creates a mapping between an already existing data source (relational database) and an SDMX data structure definition
SEA SDMX Editor	nextSoft GmbH	Simple SMDX-compliant solution for managing and maintaining statistical metadata

Working with SDMX Data Files

SDMX Converter	Eurostat	Translates between different SDMX formats or into other file formats
Fusion Transformer	Metadata Technology	Translates between different SDMX formats. Based on the *data-streaming* concept, and can therefore handle data files of any size
SDMX Transformation Component Applying CSPA	OECD	Another translation tool for SDMX files, complies with the CSPA (Common Statistical Production Architecture) framework developed by the international statistical community for the software development of statistics production processes
SDMX Java Suite	ECB	Java-based tool for reading and validating SDMX files, contains additional modules for subsequent storage in FAME databases

Working with the Data in SDMX Data Sets

Flex-CB Visualisation Framework	ECB	Toolbox for the graphical or tabular representation of SDMX-modeled data and metadata
rsdmx	Emmanuel Blondel	Collection of classes and methods for working with SDMX data sets in the statistics software R. Using rsdmx, SDMX data sets can be imported to R. Alternatively, using SDMX Web Services, rsdmx can also directly access data collections from various providers
SDMX Connectors for Statistical Software	Banca d'Italia	Components for the use of SDMX Web Services to import SDMX data into statistics standard software such as R, MATLAB or SAS

Storing information about SDMX data centrally

Euro SDMX Registry	Eurostat	A concrete realisation of the SDMX Registry described in Sect. 12.5 for the central storage and administration of SDMX data models
Fusion Registry	Metadata Technology	SDMX Registry implementation which supports all common versions of the SDMX standard. The product forms the basis for the SDMX Global Registry operated by the SDMX sponsor organisations. Available as a free community edition and as a commercial enterprise edition
SDMX Global Registry	SDMX sponsor organisations	Another concrete realisation of the SDMX registry. Currently contained data structure definitions: Balance of Payments and International Investment Position, Foreign Direct Investments, Government Finance, National Accounts

Doing it yourself: Programming libraries for implementing SDMX products

OpenSDMX	UN Food and Agriculture Organisation (UNFAO)	Components for the Java-based development of SDMX software, especially for the use of SDMX Web Services
pandaSDMX	Stefan L. Pankoke	Python-based library for working with SDMX data sets
SDMX-Experiments	James Gardner	Toolbox for working with SDMX files and services in Java and JavaScript
SDMX.NET	UNESCO Institute for Statistics (UIS)	SDMX Framework for programming on the Microsoft.net platform
SDMX Reference Infrastructure (SDMX-RI)	Eurostat	A collection of services from the SDMX world for use and further development within your own IT and data landscape
SDMX Istat Framework	ISTAT (Italian National Statistical Office)	Package of SDMX products for direct use and development, written in C#. Contains the most important building blocks for an SDMX-based data "assembly line" (Metadata Repository/Registry, SDMX Web Service, Metadata Web GUI, SDMX Builder & Loader, Data Web Browser)
SDMXSource	Metadata Technology	A container for various open source libraries for SDMX development, some of which were already mentioned
SDMX in Eviews	Louis de Charsonville	Interface to load SDMX data into Eviews software
SDMXUSE	Sébastien Fontenay	Interface to load SDMX data into Stata software

Reference

SDMX (2018) Software tools for SDMX implementers and developers. https://sdmx.org/?page_id=4500. Accessed 17 Apr 2018

Chapter 14
SDMX as a Key Success Factor for Data Integration

Abstract Data integration was already a major challenge for official statistics by the 1990s. The task at that time consisted of professionally harmonising the different national business phenomena and transferring the harmonised data sets via a file-based data exchange process into a uniform database. This was achieved with SDMX (Statistical Data and Metadata Exchange), but SDMX is much more. It is a non-technical model to classify any data world and thus come to a uniform view and approach to its data. Using SDMX, it was possible to build very extensive data collections on a variety of topics.

It is, therefore, not worth waiting for a better standard. Since standards draw their strength from their dissemination and less from their genius, this would be futile. It is important to recognise the power in a potential standard, and then expand it and, above all, promote its dissemination.

Though data integration seems to be the word of the hour, it was already a major challenge in the 1990s, especially for official statistics. At that time, data integration was less driven by the sheer size of the data and the explosive expansion of the volume of data; rather, during the European integration process and in preparation for the Monetary Union, it was a question of the harmonisation of different national data worlds—and was therefore content-related. And data integration was more focused on classical data exchange processes.

This meant that different national perspectives had to be harmonised content-wise, and the resulting harmonised data sets had to be brought together via a file-based data transmission procedure to form a uniform data collection. To do this, it was decided early to focus on a time series–based data exchange.

These origins are the reason for several misconceptions about SDMX, which are characterised by the typical expressions "SDMX is for time series only" and "SDMX is only a data exchange format". SDMX is much more than that. It is a non-technical model to organise the data world and thus to arrive at a uniform way of viewing and dealing with data. Technically, it can be implemented in very different ways.

The original goals were more than satisfyingly achieved: the European institutions managed to harmonise the different data worlds and establish simple, effective data exchange processes between statistical offices, statistical units of central banks

© Springer International Publishing AG, part of Springer Nature 2018
R. Stahl, P. Staab, *Measuring the Data Universe*,
https://doi.org/10.1007/978-3-319-76989-9_14

and international organisations. This is because the interaction of data and metadata allows for the most effective expansion of data sets: every time a new domain of statistical data is to be exchanged, the professional work consists of the formulation of the harmonised statistical data set in the SDMX language. This provides the necessary metadata, which can be exchanged electronically between the parties in the form of SDMX DSDs. After that, it is clear what an actual data delivery will look like, and the data exchange process can be started without any additional program adjustments of IT systems. Following this approach it was possible to build very extensive data collections on a wide range of topics. There were no new IT projects required, only the purely content-related work and a subsequent use of the existing standards. In this way, comprehensive data collections were developed, without the necessity for large projects. Examples can be found in the ECB (SDW), at the OECD, on the websites of the Banque de France, the Bundesbank, the BIS and other institutions already mentioned.

The IMF's SDDS Plus initiative, presented in Sect. 9.5 of the book, is a fully fledged realisation of the SDMX vision. A central repository maintained by the IMF provides the metadata information for a comprehensive collection of key economic indicators relevant for financial stability, together with links to the websites of the national central banks, statistical offices and finance ministries on which the actual data are stored. SDDS Plus uses SDMX for data classification, data storage and data dissemination (Piché 2013).

These examples demonstrate that SDMX is much more than a data exchange format. It provides a professional basis for comprehensive data collections and an ideal platform for data sharing and distributed information on decentralised systems.

Concrete links to examples of this success story are the statistical data portals of the ECB (http://sdw.ecb.int) and the OECD (http://stats.oecd.org), as well as two IMF-operated sites: the "Principal Global Indicators" (http://www.principalglobalindicators.org) published by the G20 Data Gaps Initiative and the central entrance page of the Special Data Dissemination Standard Plus, the "Dissemination Standards Bulletin Board" (http://dsbb.imf.org).

The two main strengths of SDMX are the multi-dimensionality and the interaction of data and metadata. The multi-dimensional structure makes it possible to organise and analyse data using defining characteristics—the dimensions. Given the fact that any number of dimensions is theoretically possible, any desired phenomena can be described. The interaction of data and metadata makes it easy to automate data-sharing processes, to build self-explanatory data structures and central data repositories, and thus create a map of the data universe.

Is it worth to wait for a better standard? No.

No, another standard for the generic classification of data would only be very similar to SDMX—possibly "a bit better" here or there, depending on the technical or business-related point of view. But it would be irrelevant, since standards draw their strength more from their publicity and less from their geniality. Most important is to recognise the potential in a standard, to expand it and, above all, to promote its dissemination.

Reference

Piché R (2013) The benefits of SDMX for SDDS Plus. http://www.oecd.org/std/SDMX%202013%20Session%203.11%20-%20The%20benefits%20of%20SDMX%20for%20SDDS%20Plus.pdf. Accessed 20 Feb 2017

About the Authors

Photograph copyright of the author

Reinhold Stahl, mathematician, has worked at the Statistics Directorate of the Deutsche Bundesbank since 1985. He was responsible for the creation of the Bundesbank's statistical information system in its current form, before becoming the Director General Statistics in 2014. He has been actively involved in the international success story of the SDMX standard presented in this book since its beginning and has introduced this standard into the statistics of the Deutsche Bundesbank. The opportunities opened up by the standardisation have made him a passionate advocate of this approach.

Photograph copyright of the author

Dr Patricia Staab, mathematician, started working at the Statistics Directorate of the Deutsche Bundesbank in 2000 and immediately took part in creating the internal statistical information system based on the SDMX standard. Since then, she has been appointed head of the Bundesbank's Statistical Information Management division. Both the standard and the information system based on it have developed substantially since the beginning, but the effects the standardisation could deliver at that time left a lasting impression on her.

Glossary

BI Business Intelligence: collective term for IT procedures for the systematic collection, analysis and visualisation of data

CIO Chief Information Officer: common title for the manager of the information and communication units of a company

CSV Comma Separated Values: flat file format for exchange of tabular data

Data Silo The self-contained application landscape corresponding to a specific business unit

Data Cube Form of representation for data sets, modelling them as data points in a multi-dimensional coordinate system

DWH Data Warehouse: central data collection used for analysis, into which data from different, often non-harmonised, sources are integrated

ESCB European System of Central Bank: consists of the European Central Bank (ECB) and the national central banks (NCBs) of the member states of the European Union (EU)

ETL Extract–Transform–Load: common term for all procedures designed to load data from different sources into a central unit, for example, a data warehouse

FAME Forecasting Analysis and Modeling Environment: a database optimised for time series data, from SunGard

Hadoop A Java-based framework from Apache, designed for scalable distributed IT software

HTML Hypertext Markup Language: markup language used for designing web pages

ISO International Organisation for Standardisation: promotes worldwide standards used for industry and commerce

JSON Javascript Object Notation: very compact and simple data exchange format

LaTex Text editor, commonly used in academic and scientific fields

Machine Learning Process of training an IT algorithm using existing or sample data

Markup Language Language which uses special characters to structure text-based information

© Springer International Publishing AG, part of Springer Nature 2018   113
R. Stahl, P. Staab, *Measuring the Data Universe*,
https://doi.org/10.1007/978-3-319-76989-9

OLAP Online Analytical Processing: high-performance IT methods for interactive data analysis

Open Source Software whose source code is public and therefore can be viewed, modified and used

SDMX Statistical Data and Metadata Exchange: international statistics standard, used for data classification and data exchange

SGML Standard Generalized Markup Language: markup language from the 1990s, used only moderately in industry

Web Services Concept for sharing information or instructions between IT programs, when they run on different platforms

XML eXtensible Markup Language: markup language on which countless derivatives—derived file formats—used in industry and commerce are based

Index

Printed in the United States
By Bookmasters